Stat Shot

A FAN'S GUIDE TO HOCKEY ANALYTICS

ROB VOLLMAN

Published by ECW Press
665 Gerrard Street East
Toronto, ON M4M 1Y2
416-694-3348 / info@ecwpress.com

Editor for the press: Michael Holmes
Cover design: Michel Vrana
Cover image: Ryan Stimson
Cartoon illustrations: Joshua Smith

LIBRARY AND ARCHIVES CANADA
CATALOGUING IN PUBLICATION

Vollman, Rob, author
A fan's guide to hockey analytics / Rob Vollman.

(Stat shot ; 2)
Issued in print and electronic formats.
ISBN 978-1-77041-412-9 (softcover)
ISBN 978-1-77305-250-2 (HTML)
ISBN 978-1-77305-251-9 (PDF)

I. Hockey—Statistics. 2. Hockey—Statistical
methods. I. Title.

GV847.V64 2018   796.96202'1   C2018-902547-6
C2018-902548-4

The publication of *Stat Shot: A Fan's Guide to Hockey Analytics* has been generously supported by the
Government of Canada. *Ce livre est financé en partie par le gouvernement du Canada.* We also acknowledge
the contribution of the Government of Ontario through the Ontario Book Publishing Tax Credit and the
Ontario Media Development Corporation.

At ECW Press, we want you to enjoy this book in whatever format
you like, whenever you like. Leave your print book at home and take
the eBook to go! Purchase the print edition and receive the eBook free.
Just send an email to ebook@ecwpress.com and include:

**Get the
eBook free!***
*proof of purchase
required

• the book title
• the name of the store where you purchased it
• your receipt number
• your preference of file type: PDF or ePub

A real person will respond to your email with your eBook attached.
And thanks for supporting an independently owned Canadian publish-
er with your purchase!

FOREWORD

by CRAIG CUSTANCE

NHL insider, editor-in-chief of The Athletic Detroit, *and author of*
Behind the Bench: Inside the Minds of Hockey's Greatest Coaches
Detroit, Michigan

It was a room full of people at one of the largest analytics conferences in the world. On stage, there was a nice mix of traditional hockey voices and those knee-deep in analytics. Everything was in place for an informative session on hockey analytics and how they should best be used, and for a good 10, maybe 15 minutes that's exactly what the audience received.

But slowly, as the session continued, the conversation strayed from analytics and into a general hockey question-and-answer session. When a panellist started talking about the benefits of the larger European ice versus North American ice, my hope of getting a substantial lesson in analytics was gone. Years ago, this was how it often went.

There's a reason for that. Talking about analytics in a way a large audience can easily consume while still being in-depth enough to satisfy those with more than a passing knowledge of the subject is really hard.

Start explaining the calculations for Euclidean distance and how it factors into your analysis, and you're probably going to lose any fan who just wants an understanding of hockey analytics beyond Corsi.

The same thing happens when you present a chart that looks like somebody puked NHL team logos onto a spiderweb.

But in the last couple years, we've also seen an overreliance on basic shot-based statistics sold as advanced metrics. I've been guilty of it myself. Hockey fans now want context and a deeper-level analysis, beyond suggesting a defenceman is great because he has a 52.3% Corsi for.

Rob Vollman always hits the sweet spot.

I got to know Vollman when we were working together at ESPN, when hockey analytics were just emerging from the obscure writings of some of the original ground-breakers into everyday conversations. What was new to many of us traditional hockey writers on ESPN's staff was stuff Vollman had been working on for years, and so he became the guy we went to with questions.

During an era in which there was often a rift between the traditional media and those developing hockey analytics, Vollman was exactly the opposite. He was patient in answering questions and emails that came from our hockey-writing group as we tried to grasp the different concepts.

Soon, in part because of his insight, we were working hockey analytics into stories, helping round out our coverage. While some corners of the media were hostile and resisted the analytics movement, we worked hard to figure out where it fit into traditional hockey coverage. Having Rob in those discussions, quick to answer any and all questions in a way we all understood, made the process anything but hostile. It was educational. It was fun. It made us better writers.

It's all the things that make reading his books worthwhile.

There's a reason people keep buying the Hockey Abstract series, like we all used to buy season preview magazines from the newsstand. And in this sequel to *Stat Shot*, readers are treated to the same mix of clear analysis, a little humour, and complex ideas presented in such a way that the pages keep turning.

It's been fascinating to watch the adoption of hockey analytics come in fits and starts. It hasn't been a linear process, but when you pan out, you see that the progress has been constant. For years now, Vollman has been a part of that constant march toward mainstream acceptance of hockey analytics. His consistent, clear guidance has made us all better viewers of the game.

If this is your first time reading Vollman on analytics, I'm excited for what you have ahead. You're going to laugh, you're going to question how you watch hockey, and, most importantly, you're going to become better equipped to soundly analyze hockey. If this isn't your first time reading Vollman, you know what you're in for. Enjoy!

INTRODUCTION

"Go ahead, give him another one," my older brother Mike smilingly prodded his friend.

"Anyone?"

"Yes, anyone. Any page, any player," my brother insisted.

"Okay." My brother's friend flipped through my well-worn copy of the *1988 Baseball Register* and studied the pages, brows furrowed. I watched on nervously, but my brother had a relaxed and confident grin on his face. "Okay, got one," his friend said at last. "Dickie Noles."

I breathed a sigh of relief because I knew this player well. "Dickie Noles, played with the Chicago Cubs and Detroit Tigers last year," I reported from memory. "He was 4-2 with an ERA around 3.53."

Mike smiled at his friend, who nodded to confirm my information was correct.

"That was actually his best ERA ever. I think it started in the 3.80s in his first two seasons with the Phillies, but really shot up after that," I continued. "His best season was around 1982, with the Cubs, when he was 10-13. Other than that, I don't think he ever won more than five games in a season."

By this stage, my brother was chortling at the shocked look on his friend's face. "So, he memorized everybody's stats?" his friend asked. "Is there something wrong with him?"

As a child, I often wondered if there was something wrong with

– FROM ICE LEVEL

me, and my love of numbers. Sure, it was a source of amusement for my brother and very convenient for my parents to have a walking calculator when a nation-wide 7% sales tax was introduced in 1991, but none of my friends seemed to share my passion for numbers.

That's why the discovery of the *Bill James Baseball Abstract 1984* was such an important event in my life. Not only had I found someone who apparently looked at the world the same way I did, but he showed me the tremendous benefits of doing so.

It didn't take me long to start applying these ideas to my favourite sport, hockey. I even discovered and fully digested a few of the early books on hockey analytics, like Stan and Shirley Fischler's *Breakaway '86*, and the *Klein and Reif Hockey Compendium*. I never dreamed that one day I'd be publishing an analytics book of my very own, but that day eventually came.

After several seasons of co-authoring annual guide books in the Hockey Prospectus series, I finally published my own book in 2013, *Hockey Abstract*. It was obviously named after the Bill James books from my childhood, and I even designed the structure and layout to match (please don't sue). It was organized into 10 chapters that asked questions in true James-like fashion, followed by 10 chapters that provided some background on key statistics and related principles—10 being such a nice, round number.

The book was a success, not just in terms of sales but in terms of getting people hooked on the world of hockey analytics, much like Bill James had inspired me so many years before.

My fellow fans weren't the only ones paying attention, and I soon found myself in an NHL GM's office. He was absolutely fascinated by what I had explored in those pages, and we swapped stories and opinions for hours. I hadn't planned on writing another book, but by the time I left his office, he had convinced me of the importance of this work.

I set about writing a sequel, imaginatively entitled *Hockey Abstract 2014*. I found more questions to explore, included two-page, James-like essays for each team, and added contributions from two of my earliest mentors in this field, Tom Awad and Iain Fyffe.

Once I'd written the book, it was time to spread the word. With the encouragement of family and friends, I went deep into my own pocket and launched it at a hockey analytics conference that I'd organized at the Pengrowth Saddledome in Calgary, Alberta, on September 13, 2014. The support this event received was highly rewarding. Chris Snow of the Calgary Flames was my opening speaker, local talk show host Rob Kerr of Sportsnet FAN 960 was my master of ceremonies, and Andrew Thomas of Carnegie Mellon and Dr. Michael Schuckers of St. Lawrence University spoke. All four volunteered their services and were among the many good friends I made that day.

Thomas and Schuckers went on to host their own hockey analytics conferences. Similar events have now been held by various people in eight different cities, including regular events in Ottawa, Rochester, and Vancouver. It was at Dr. Schuckers's first event at Carleton University in Ottawa, with his co-host Dr. Shirley Mills, that I received my greatest reward. You might assume that the aforementioned meeting with an NHL GM was the highlight of this journey, but it was yet to come. Flanked by Tom Awad, who had driven up from Montreal to attend the conference with me, I was approached by a number of young men and women holding copies of my books and a pen. One by one, they explained to us how our work had inspired them, using much the same language that I would have used to describe Bill James's work.

The success of these two books eventually led, in 2016, to *Stat Shot*. The opportunity to reach a wider audience inspired me to add some more ambitious content. The opening chapter, for example, was a blueprint for a team management model. It wasn't easy to make topics like weighted averages, regression, aging curves, and the NHL's collective bargaining agreement fun and accessible, but it certainly sounds like we hit the mark. Plus, Joshua Smith's cartoons helped keep things light.

Given the success of *Stat Shot*, which sold out its first printing in short order, was ranked as the number one hockey book on Amazon for months, and even made an appearance on the *Globe and Mail* best-seller list, plans for a second book were quickly put in motion.

With *Hockey Analytics for Everyone*, the big news is that we'll be venturing outside the NHL for the first time. After all, it's meant to be a guide to hockey analytics, not just NHL analytics.

This progression is long overdue. During the 2005 lockout, I had a Swedish friend tease me about how I must be struggling to cope with life without hockey. Living in Calgary, a city that hosts the WHL's Hitmen, the University of Calgary Dinos in the CIS, the women's national team, a great Midget AAA tournament, and, of course, my Friday night recreational league, it wasn't exactly life without hockey. The NHL may be hockey's biggest and best show-case, but it's hardly the only one.

That's why it's great to finally expand the Hockey Abstract series outside the NHL. This book has a chapter dedicated to the hunt for the world's best women's hockey player and a chapter exploring how to translate data from other leagues, and a chapter about how to place hockey stats in context includes some honorable mentions of other leagues. Plus, in an effort to cover more of what hockey represents, there's even a chapter that goes beyond the players and focuses on the coaches.

I'm also excited about the opening chapter, which outlines the absolute basics of hockey statistics. It's fair to wonder why an intro-ductory chapter like that wasn't included in *Stat Shot*. When I began writing it, I didn't know how it would be marketed or even what we'd call it. I wrote it simply as the third installment in the Hockey Abstract series, picking up where the others had left off. I certainly didn't know it would also have a subtitle that claimed it was the *ultimate* guide to hockey analytics! Nobody complained about that audaciousness, but, in fairness, it also happened to be the *only* guide to hockey analytics.

Hockey Analytics for Everyone allows me to correct that oversight and kick things off with an introductory chapter that starts from absolute zero. There's no Corsi, no PDO, and no math besides simple multiplication and division. And yet, this chapter provides all of the basics you need to know to enjoy not only the rest of this book, but virtually everything else in this field.

At the other end of the spectrum, the most ambitious content in this book attempts to build a career projection model for goalies in order to find the NHL's most valuable one. It was a real eye-opener for me, especially how the world's best goalie, Carey Price, only barely makes the top 10 once we take factors like age and cap hit into account.

The book closes with a project I've been working on for years: a comprehensive glossary of every hockey statistic. More than just a few simple sentences about Corsi and PDO, I traced the origins, meaning, and exact formula for every stat from the early days of *Breakaway '86* and *The Hockey Compendium* to the leading-edge hobbyist websites of today. I'm hoping that this is just one more way to inspire more exploration and create a book that will remain on your shelf as a timeless reference.

HOCKEY STATS 101

Hockey stats really aren't that difficult—once you break things down.

One of the reasons I rarely use the term "advanced stats" is that there's really nothing terribly advanced about what hockey statisticians do. Everything starts with a simple counting statistic, then we account for opportunity, and then we place the data in context. As I hope to demonstrate, this is a simple three-step process that is easy to grasp, even for those without a mathematical background.

Hockey stats are at their best when they serve as a sober second thought and help point out things that we missed. After all, it's easy for our eyes to deceive us. We get so swept up in the emotion of the moment—and we all have our biases about teams and players—that we sometimes don't really see what's happening on the ice. Even when we see the game clearly and objectively, we rarely remember the important details the next day.

However, without a proper understanding of how to use them, stats can be just as deceiving as the perspective of the most emotional and biased fan. Just as in any other field, we can only achieve a clear interpretation of hockey statistics by taking clearly defined and accurate measurements, adjusting those measurements for opportunity, and placing them in context. Even if you choose to skip this chapter, understanding that means you have understood the essence of hockey analytics.

There are so many simple examples of that clear interpretation, that there's no need to look at any stats with fancy names, like Corsi, Fenwick, and PDO. In this chapter, we stick to simple stats like goals and wins. These are excellent base statistics. Since everyone understands and uses them, they have a clear and universally accepted definition and their importance is obvious.

Following those base stats, we'll explore how to take opportunity into account, by calculating stats like goals per 60 minutes and winning percentage. The third and final step is to place that information in context by using charts, rankings, and comparisons to the league average. Finally, for the particularly ambitious, we'll close by introducing goals created, which is a compound statistic meant to replace points.

Team Stats

There is no better place to start than with wins. It is the entire point of hockey and a concept that everybody understands. It's the only stat that truly matters, and everything in the world of hockey analytics either boils down to wins or is utterly meaningless.

So let's start with wins. Better yet, let's start with 36 wins. What does 36 wins mean, other than a team outscored its opponents in 36 separate games? Quite frankly, it doesn't mean much.

Wins may be the ultimate statistic, but they mean nothing without opportunity. For example, if we're studying the Chicago Blackhawks, who won 36 games in the 2012–13 season, which was shortened to 48 games because of a lockout, then 36 wins is an incredible achievement. It means that the Blackhawks were one of the most dominant teams in NHL history. However, if we're talking about the 36 wins by the Vancouver Canucks the following season, which was over an 82-game schedule, then it doesn't mean quite so much.[1]

That's why the only truly important statistical adjustment accounts for opportunity. In this case, the number of games a team plays represents the number of opportunities that it had to win.

---

1. Chicago and Vancouver team statistics from NHL.com - Stats, National Hockey League, updated July 7, 2017, https://www.nhl.com.

Chicago had 48 opportunities, and Vancouver had 82. Dividing 36 wins by the number of games produces each team's winning percentage. For Chicago it's 0.750, meaning Chicago earned 0.750 wins per game. Given that you can either earn 0 wins or 1 win in a game, that also means that Chicago had a 75.0% chance of winning any given game. For Vancouver, it was 43.9%. That's a big difference.

Besides percentages, the other way to account for opportunity is to calculate the *rate* at which teams accumulated wins. For example, Chicago had a rate of one win every 1.33 games, which is 48 games divided by 36 wins, and Vancouver had a rate of one win in every 2.28 games.

You will generally never see a team's winning rate presented in those terms, since it's not easy for us to place such numbers in context. When presented with no other information, it's hard to accurately figure out exactly how good 1 win in every 1.33 games actually is. After all, teams play one game or two games but never 1.33 games. (Well, maybe the old Toronto Maple Leafs would sometimes play only 0.33 games in a night, but things have changed.)

As fans, we think in terms of an 82-game schedule, so it's better to present a team's winning rate in those terms: Chicago had a rate of 61.5 wins per 82 games, while Vancouver obviously had a rate of 36 wins in 82 games. That statistic is much easier to place in context, and it helps illustrate the importance of finding the right terms in which to express a rate statistic.

We now need to take a step back because I actually wrote a little bit of a fib at the top of this section. Wins may be the ultimate statistic in the playoffs, but they don't actually hold that distinction in the regular season, where points are king. As it turns out, teams can earn points not only from winning but also from ties (which existed prior to 2005–06) and even for certain types of losses (since 1999–00).

That means that a team can make the playoffs despite winning fewer games than another. In fact, it happens all the time. Florida won 42 games in 2016–17, but Toronto, who had 40 wins, made the playoffs.[2] So wins are *not* the ultimate statistic. Sorry about that.

---

2. Florida and Toronto team statistics from NHL.com - Stats, National Hockey League, updated July 7, 2017, https://www.nhl.com.

-CHARA VS. GAUDREAU

TWO MINUTES FOR ~~HIGH~~ LOW STICKING!

This is why the NHL doesn't actually use winning percentage anymore, but rather points percentage. Points percentage works the same way as winning percentage, by dividing a team's points by their opportunity to earn points. Since a team can earn up to two points per game, their actual points are divided by the maximum number of points they had the opportunity to earn, which is the number of games they played multiplied by two.

Continuing the example, Chicago earned 77 points in 48 games, and 77 divided by 96 (which is two points per game over 48 games) works out to a points percentage of 0.802. That means Chicago earned an average of 0.802 multiplied by the maximum two points per game, which equals 1.604 total points per game. However, since teams can earn zero, one, or two points per game, it no longer means that they have a 80.2% chance of earning a point in any given game.

Rates are also calculated the same way for points as they were for wins. In this case, Chicago earned points at the rate of 131.5 points per 82 games. That's a much more meaningful number than 0.802.

These are admittedly very simple concepts, but the principles of taking opportunity into account and calculating rates gets trickier when we apply them to other statistics and other situations, so it's important to get a good grasp of them up front.

Before moving on to how to put these numbers in context, let's step back to the core of hockey analytics: counting statistics. To explore this, we'll look at goals, which are the source of wins. As previously established, teams win games by scoring more goals than their opponent. That's why there is no closer relationship between any two hockey statistics than the one between goals and wins.

Goals are the best example of a counting statistic, which is exactly what it sounds like—it's anything that can be counted. If you're watching a hockey game and can point at an event and say "Hey look, there's one and there's another one and there's another," then it's a counting statistic. Needless to say, there are an almost endless number of counting statistics at your disposal in any given game, like goals, shots, hits, penalties, passes, tears shed by the Leafs fan next to you, and so on.

You can define counting statistics in many different ways, but the common thread is that they are events that either occurred completely or did not occur at all. For example, a goal occurs when the puck completely crosses the goal line before regulation time expires in the judgment of the goal judge. A team gets absolutely no credit whatsoever if it gets the puck 99.9% across the line or if the puck crosses one microsecond after time expires.

Counting statistics should also have a clear and complete definition. Continuing with goals, a goal must occur without the attacking team committing an infraction, like goalie interference, and in a legal fashion, such as without a high stick or a kicking motion, again in the judgment of the officials.

Unfortunately, you may find that many common statistics lack a clear and complete definition and are therefore subject to the whims and opinions of the scorekeeper. Be wary of any statistic based on these subjective counting stats, like hits or takeaways.

As previously discussed, to account for opportunity, counting statistics can be converted into rates. In this case, a team's goal-scoring rate can be calculated on the same 82-game basis as wins and points, but placing it in per-game terms is another useful context for the average fan. For example, Chicago scored 155 goals in 48 games, which is 3.23 goals scored per game or 264.8 over an 82-game season.[3] These numbers can be more easily understood by the average fan, with no pencil or paper required.

---

3. Chicago's team scoring statistics from NHL.com - Stats, National Hockey League, updated July 7, 2017, https://www.nhl.com.

To get technical, some statisticians like to calculate goals per 60 minutes instead of goals per game, since not every game is of the same length. Some games have up to an additional five minutes of overtime, while others do not. But we'll have plenty of time to get into the more pedantic details later in this book. For now, let's keep exploring the relationship between goals and wins.

Consider the following chart, which was taken from *Stat Shot*.[4] Each dot represents a single team's regular season. On the horizontal axis is the team's goal differential, and on the vertical axis is the number of points they earned in the standings. As you can see, there's a pretty direct relationship between goals and points, and most teams don't stray very far from the trend line.

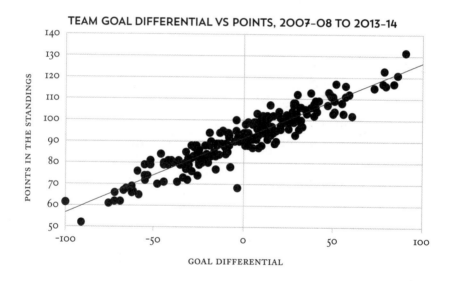

TEAM GOAL DIFFERENTIAL VS POINTS, 2007-08 TO 2013-14

Goal differential is simply the number of goals a team scored minus those it allowed. For example, Vancouver scored 196 goals in 2013–14 and allowed 223 goals, for a goal differential of −27. Having earned 83 points in the standings, the Canucks are the dot located almost exactly on the line, at -27 on the horizontal axis and 83 on the vertical.

In the larger sense, differentials are a difference within a single

4. Rob Vollman, "What's the Best Way to Build a Team?," *Hockey Abstract Presents . . . Stat Shot: The Ultimate Guide to Hockey Analytics* by Rob Vollman with Tom Awad and Iain Fyffe (Toronto: ECW Press, 2016), 39.

counting statistic. For instance, you can calculate a team's penalty differential by subtracting the penalties they took from those of their opponents. However, it doesn't make statistical sense to create a differential from two different counting statistics, like subtracting passes from shots to create a shot-pass differential. That sort of comparison usually involves creating a compound statistic, which we're not exploring just yet.

Differentials make the most sense when each side has the same opportunity to generate the counting statistic in question. That's certainly the case with goals, since both teams are on the ice at all times.

Getting back to the chart, if teams win games by outscoring their opponents, then why aren't all the dots *exactly* on the trend line? Why would they vary at all?

Since winning isn't the only way to get points, teams that earned a lot of points for losing games in overtime or the shootout are above the line. Plus, teams that won a lot of close, one-goal games but lost a few massive blowouts are also above the line. That type of data tends to even out over the long run, but not always over 82 games.

There is another limitation with differentials that can explain some of the variance on the chart. Differentials indicate how many more counting events occurred over a certain period of time, but they offer absolutely no indication of scale. For example, a goal differential of +15 over an entire season is really nothing special, but it's incredible over a single game.

For a real-life example, consider the 2009–10 Vancouver Canucks, who had a goal differential of +50, and the 2011–12 St. Louis Blues, who had a similar goal differential of +45 in the same number of games.[5] On the surface, the Canucks appear to have been slightly more effective at outscoring their opponents and ought to have finished with one or two more points. However, if we dig deeper into the numbers that make up the differential, it becomes clear that St. Louis actually had a larger share of all of the goal-scoring.

In raw terms, the Blues outscored their opponents 210-165, while Vancouver outscored their opponents 272-222. Placed in terms of a

---

5. Vancouver and St. Louis team statistics from "NHL.com - Stats," National Hockey League, updated July 7, 2017, http://www.nhl.com.

goal percentage, St. Louis was responsible for 56.0% of all the goals in its games, which is 210 divided by the sum of 210 and 165, while Vancouver was responsible for 55.1%. That may be a subtle difference, but it could be one of the reasons why St. Louis actually had more points than the Canucks, 109 to 103.

While counting statistics are sometimes presented as differentials in mainstream coverage, they are almost never offered in terms of a percentage. Unless you're a long-time reader, you have probably never seen information presented this way. In essence, this is the point where we are finally starting to pierce the skin of the world of hockey analytics and bite into the delicious fruit inside.

GOAL PERCENTAGES, 2011–12

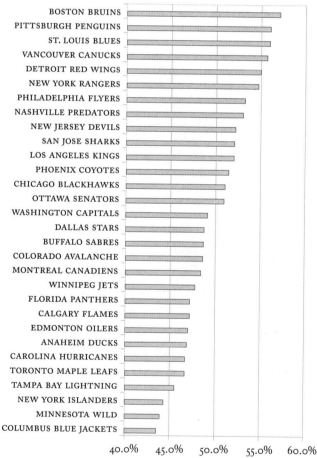

As we enter this world, the greatest challenge is how to place new and unfamiliar statistics in context. Most seasoned hockey fans have a good sense of what it means for a team to earn 109 points or to average 3.23 goals per game, but many of the metrics in this book will be new to most fans. Be honest, do you have any idea how good a goal percentage of 56.0% actually is? I mean, you know that it's above an average 50%, but is that really far above average or just a little?

That's why it's absolutely critical to find ways to place these new stats in context. The simplest way is to mention where a result ranks. In this case, the Blues ranked third (out of 30 teams) in 2011–12. Using ranks is a very effective way of placing a new or unfamiliar stat in context, especially when it's expressed visually, as follows. At a glance, you get a real sense of how good 56.0% is, since it's now relative to the rest of the league. (See page 122 to 148 for more information about how visualizations can help place individual stats in context.)

Another method of placing stats in context is to express them relative to the league average or some other point. In this case, the Blues' goal percentage of 56.0% would be +6.0% relative to the league average of 50.0%. This is commonly referred to as a relative statistic. That is, the Blue's relative goal percentage is +6.0%.

To take a trickier example, Chicago's average of 3.23 goals per game in 2012–13 is greater than the league average of 2.73 by 0.50 goals per game in absolute terms, which is 18.3%. In this case, Chicago's relative scoring rate is +0.50 goals per game, and its relative scoring percentage is +18.3%.

This technique makes it possible to compare Chicago to great teams in other leagues or from other eras, like the dynastic Edmonton Oilers, who scored 5.58 goals per game in 1983–84, the season in which they won their first Stanley Cup. The league average that season is 3.95 goals per game, meaning that Edmonton's relative scoring rate is +1.63 goals per game, and their relative scoring percentage is +41.1%.

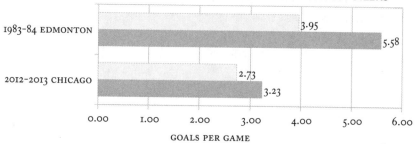

2012-13 CHICAGO BLACKHAWKS VS 1983-84 EDMONTON OILERS

GOALS PER GAME

As great as Chicago was in 2012–13, the 1983–84 Oilers were a better team, offensively. This kind of comparison is only possible when statistics are placed in context.

That covers a lot of the basics for team-based stats, and it's everything you'll need to know to enjoy the rest of this book: start with a clearly defined counting statistic, account for opportunity by calculating a rate, and then place the information in context using charts or techniques like ranks, percentages, or commonly understood formats. Now let's apply this knowledge to individual players.

Individual Player Stats

Since the earliest days of the sport, goals have been the most common statistic used to evaluate individual hockey players. Given that the meaningfulness of goals is obvious, it's a great place to start.

Every goal is awarded to the single attacking player who was deemed to be ultimately responsible for the goal. Typically, that is the player who either took the shot or last deflected it, intentionally or otherwise. In almost every case, goals are assigned to the attacking player who last touched the puck before it crossed the goal line, which also helps cover rare situations, like when a goal is accidentally scored by a defensive player or the goalie.

While the process of counting an individual player's goals scored is relatively simple and, at the team level, accounting for opportunity and placing those results in context is straightforward, it poses more of a problem at the individual level.

Team statistics are generally easier to handle than individual player statistics, since every game is a head-to-head competition. In theory, each team has an equal opportunity to perform, and everything that occurs is a direct consequence of the actions taken by one of those two performers.

At the individual level, this is no longer the case. At any given time, there are between 8 and 12 individual players on the ice, with any remaining players on the bench, in the penalty box, or removed from the game due to injury or penalization. Unlike a head-to-head matchup, not every player has an equal opportunity to perform, nor an equal influence over each of the game's events. It is therefore more challenging to measure an individual player's opportunity and to place these results in context.

At the team level, opportunity is taken into account by calculating goals as a rate. Specifically, an NHL team's goals can be represented as a rate over an entire 82-game schedule or over 60 minutes of play. The same thing can be done at the individual-player level. For instance, as of the end of the 2016–17 season, Washington's Alex Ovechkin has scored 558 goals in 921 games or 19,380 minutes played in his career,[6] which works out to 49.7 goals per 82 games and 1.73 goals per 60 minutes.

Comparing individual goal-scoring using those rates is an excellent start, but it's not perfect. After all, someone playing on the top line or on the power play with Sidney Crosby or Connor McDavid has far more opportunity to score goals than the opposing defenceman assigned to shut them down, even in an equal number of games and/or minutes played. Comparing someone in the former situation to the latter won't work, since the first player had far more opportunity to score than the second. In other words, an individual player's goal-scoring rate is only an accurate measurement of his ability to score goals if his situation is taken into account.

While this is not meant to be an exhaustive list, consider the following 10 situations in which goals are easier for individuals to score, to one extent or another:

---

6. Alex Ovechkin's career statistics from NHL.com - Stats, National Hockey League, updated July 7, 2017, https://www.nhl.com.

- When playing alongside particularly talented linemates, like Crosby or McDavid.
- In certain manpower situations, such as a five-on-four power play or three-on-three overtime.
- When playing against particularly weak opponents.
- When facing a particularly weak goaltender.
- When frequently assigned shifts that begin in the offensive zone.
- When, due to the score, the opposing team isn't in a defensive shell.
- When the opposing goalie has been pulled for an extra attacker.
- When assigned a penalty shot.
- When playing at home.
- In the second period, when teams switch ends and players have to go farther to complete a line change.

This is exactly how and why statistical hockey analysis can get complicated. Even with a statistic as simple and as commonly understood as goals, it can be very difficult to accurately compare one player to another because of all the different situations that can occur. But please don't get overwhelmed. Regardless of the individual statistic being studied, or the situations that apply, there are only four basic ways to deal with this problem.

▄▌ 1. DO NOTHING

The first option is to do nothing (it's not as bad as it sounds). There's nothing inherently wrong with this approach, provided that the factors in question have a relatively minor influence over scoring (although this is not the case here) or that the audience is aware that the results can't be used to compare different types of players.

For example, even if we do nothing to adjust their goal-scoring rates, we can still compare Ovechkin to someone else who is used the same way, on the top line at even strength and on the power play, such as Steven Stamkos.

Ovechkin outscores Stamkos 558 to 321, but that's because he has played more. Stamkos has scored those 321 goals in 586 games or 11,546 minutes, which works out to 44.9 goals per 82 games and

1.67 goals per 60 minutes.[7] At that rate, it would take Stamkos 20,070 minutes to score 558 goals, while Ovechkin has played only 19,380 minutes. When placed in context, that extra 690 minutes looks like an awfully small difference, so it's hard to be certain that Ovechkin is truly better.

This is a very insightful way to compare these two players, but we also know that this comparison isn't exact, since they play for different teams. That means that a lot of factors could easily account for such a small difference in their scoring rates, like their linemates or coaching strategies. Let's pick just one for further exploration: manpower situation.

▌▌▌ 2. SPLIT IT UP

To account for differences in manpower situation, the easiest option (other than doing nothing) is to break the information down into each individual situation and then examine them separately. So rather than comparing a player's overall goal-scoring rate, we can measure and calculate it separately at even strength, on the power play, and when short-handed. Then we can compare players in the same manpower situation. In the world of hockey analytics, it is very common for teams and players to be compared in even-strength situations only.

It doesn't help us very much in this particular example, however.

---

7. Steven Stamkos's career statistics from NHL.com - Stats, National Hockey League, updated July 7, 2017, https://www.nhl.com.

Stamkos has scored 3 goals in 337 minutes short-handed, 207 goals in 8,980 minutes at even strength, and 111 goals in 2,229 minutes on the power play. That works out to goal-scoring rates of 0.53, 1.38, and 2.99 per 60 minutes, respectively. For Ovechkin, similar calculations work out to 1.28, 1.39, and 2.85 goals scored per 60 minutes.

While this approach would work with most players, it's still too close to call between these two. They're about the same at even strength, and Stamkos has the better result on the power play. Ovechkin may have a much higher goal-scoring rate when short-handed, but neither player is used very much in such situations. With just five more short-handed goals, Stamkos would have the better rate. We need a different way to express this information.

■■I 3. RECOMBINATION

Since there's an obvious relationship between how well a player scores in one situation versus another, this leads to the third option, which is to uniformly recombine these components.

As shown in the following chart, Stamkos and Ovechkin have spent a combined average of 2% of their time playing short-handed, 77% of their time playing at even strength, and 21% of their time on the power play.

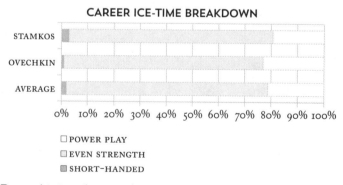

CAREER ICE-TIME BREAKDOWN

□ POWER PLAY
□ EVEN STRENGTH
■ SHORT-HANDED

Recombining their goal-scoring rates in those relative amounts, both players average 1.70 goals per 60 minutes. If we add another digit, then Stamkos has the edge, 1.706 to 1.699.

The recombination approach is only an option here because Stamkos and Ovechkin are used in roughly the same way. It wouldn't work with a player who rarely works the power play or with someone who spends a lot of time killing penalties because there wouldn't be adequate data in certain manpower situations, which would skew the recombined totals. That's why this third option is quite uncommon.

■■▮ 4. CALCULATE ADJUSTMENTS

That leads to the fourth option, which is to calculate the impact of each situation on what's being measured and adjust accordingly. This is what's happening behind the scenes of every stat with "adjusted" in its name, like adjusted save percentage for goalies or score-adjusted Corsi for teams (I know I said there would be no Corsi in this chapter, but I was just listing it as an example, so that doesn't count).

Let's stick with our ongoing example with an individual player's goals. Overall, goal-scoring rates are 2.55 times higher at even strength than when short-handed, and they are 2.86 times higher still on the power play.[8]

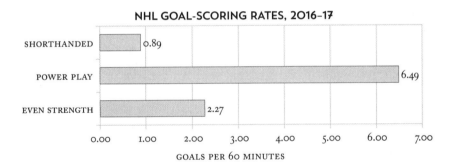

NHL GOAL-SCORING RATES, 2016–17

SHORTHANDED 0.89
POWER PLAY 6.49
EVEN STRENGTH 2.27

GOALS PER 60 MINUTES

To compare Stamkos to Ovechkin, we could multiply each player's goals by the difficulty of scoring in the given manpower situation. In other words, we multiply their short-handed goals by 2.55 and divide their power play goals by 2.86 in order to create an equivalent of even-strength goals.

Sometimes these multipliers are called translation factors, as we translate goals from one manpower situation to the equivalent in

---

8. This raw data for my goal-scoring rate estimates in different manpower situations is from "Natural Stat Trick," Natural Stat Trick, accessed July 2, 2017, http://www.naturalstattrick.com.

another. (In the chapter "How Can a Player's Stats Be Compared Between Leagues?," we look at how to create translation factors to convert a player's data from another league to the equivalent in the NHL.) In this case, we multiply Stamkos's 3 short-handed goals by 2.55, add that total to his even-strength total of 207, divide his 111 power-play goals by 2.86, and then add that result, yielding a total equivalent of 256.1 even-strength goals. (Since I'm not rounding my multipliers at two digits, you may not obtain exactly the same result.) As a rate, that works out to 1.33 goals per 60 minutes. Unsurprisingly, repeating the same math for Ovechkin yields the same result. (That first option of doing nothing is looking better every day.)

In theory, you can make these kinds of adjustments for all of the factors that influence a player's goal-scoring rate, not just manpower situation. Even for someone who plays on a line with Crosby or McDavid, you can calculate and adjust for the boosting effect of these players in roughly the same way. Indeed, the competitive advantage one statistical perspective has over another is how well they identify, measure, and adjust for these factors.

Which of these four options should we select? I'll give you the same answer that I give when I'm asked if I wear boxers or briefs: Depends. It all depends on the impact the identified factors can have on our measurement, the data that we have available, and how we intend to use this statistic. For some factors, such as higher second-period scoring, we may decide to do nothing, since the impact is minor and most players get the same percentage of their total ice time in the second period. For manpower situations, like in our example with Stamkos and Ovechkin, we may go all the way to the fourth option, since it is a very significant factor and the data we need to make the adjustments is readily available. However, there's really no right and wrong answer, only choices.

We're now a bit beyond basic hockey stats, so I'll stop there (and you're definitely excused if your brain melted a little, and if you're a Leafs fan, please understand that poking fun at you was the only way to get all the other fans to sit through all this math).

We're now on the third and final step, which is to place this new information in context. Tragically, statisticians often skip this step, which could be the most important one. If you told someone that you did some math and can show that Stamkos and Ovechkin were both a 1.33 in equivalent even-strength goal-scoring and left it at that, they would leave the conversation with no more information than they had when it started (beyond the fact that you are a weird person who reads too many of Rob Vollman's books, assuming such a thing is even possible).

For our work to be valuable, it needs to be presented in a way that makes sense to everybody. For example, at an average of 20 minutes per game, 1.33 goals per 60 minutes works out to 36.4 adjusted goals per 82 games. That's something that an average fan can understand. They may be curious about what you mean by adjusted, but you can easily explain what you adjusted (and if that doesn't work, try making fun of the Leafs).

Choosing a meaningful rate may be a good start, but it's not always enough for the average fan to place the information in context. For example, how does 36.4 adjusted goals in 82 games compare to the average rate for a forward that season? Where do these players rank on their teams or in the league overall? Can you put it in a chart? What is that relative to the league average? Was this in the NHL or the AHL? And so on.

Thankfully, we can use the same techniques that we use to place team statistics in context with individual players, too. (There are far more creative and sophisticated ways of placing information in context, as we will explore in a dedicated chapter, but that certainly covers the basics—and then some.)

That's basically it. In a nutshell, we have explored how every fancy statistic is constructed. Yes, there are complicated stats that adjust for a great many factors and there are some clever ways to place them in context, but we follow the same three steps every time.

First, carefully define what you are measuring, which in this case was an individual player's ability to score goals, and then select a counting statistic that will form the basis of your measurement. In this case we used the standard definition of goals, but in other cases you can choose to craft a new one that includes shots (making it a compound statistic) or that modifies goals in some fashion, such as removing empty-net goals, penalty shots, and defensive own goals or one that assigns only partial credit for deflections.

Second, identify all of the factors that could influence the measurement (in this case, those that affect the opportunity to score goals), and then choose from the four major options available to deal with each factor.

Third, choose an effective way of placing the results in context. Goals are just a raw count, but statistical analysis allows us to use this raw data to do something useful, which is to measure a player's ability to produce goals.

And that's really it. One, two, three. Nothing could be simpler! Right?

Okay, so hockey stats can get a little overwhelming at times, but if you got this far, then you have passed Hockey Stats 101 and are ready to graduate. The basic class is dismissed, but we're going to cover a useful compound statistic for the teacher's pets who want to stick around for extra credit.

Goals Created

Aside from goals, the statistic most commonly used to evaluate individual players is assists.

We've already explored the definition of a goal, but assists are a little trickier to define. It's a highly subjective statistic that's based on which players contributed to the goal, most commonly with a pass or a shot of their own. Two scorekeepers can watch the exact same play and arrive at a different conclusion about which players should be awarded an assist.

Goals and assists are combined to create points, which is seen as the ultimate individual player statistic. However, combining goals and assists is an unusual thing to do, when you think about it. Goals and assists are very different things. A goal is almost always the result of an attacker's attempt to score, but an assist can be the result of an attempt to score, a direct pass before the shot, a pass to someone responsible for one of those two events, a breakout pass in the defensive zone, or even just some incidental contact shortly before the goal was scored. Furthermore, there could be a player whose actions were critical to the goal being scored who received no assist at all, like someone who threw a hit to create the turnover, someone who set up a screen, or someone who created a key diversion similar to Mario Lemieux's famous fake shot from the 2002 Olympics in Salt Lake City.

Adding together goals and assists seems almost as arbitrary as adding together any other two kinds of events, like goals and the faceoff wins that preceded them. However, since no single statistic can capture a player's contributions to a team's scoring, a compound statistic like points is clearly required. But is there a better way to combine them?

Assists require goals, but goals do not require assists. There is exactly one goal scorer per goal, but there can be zero, one, or two assists per goal. (In fact, prior to the 1930s, there was the rare possibility of three or even four assists per goal.) And so, you can combine goals and assists to assess their relative importance to creating goals. Unassisted goals could count for a full point, a goal with one assist could count as a half-point for each player involved, and a goal with two assists could count as a half-point for the goal scorer and two quarter-points for the players who assisted. Instead of having to go through every single goal and assign different values to identical events, we could just assign 0.5 points for a goal and 0.3 for an assist, depending on the average number of assists per goal that season, which is 1.68 in 2016-17. Since the goal is worth 0.5 points, the remaining 0.5 is shared by (divided by) the 1.68 assists, which works out to 0.3 for each assist.

That brings us to goals created (GC), which is a compound statistic meant to replace points that is calculated exactly as we have devised. A particularly nice benefit of this formula is that the individual GC of every player on an entire team adds up to the total number of goals that team scored.

The difference between GC and points may be subtle, but it's easy to see how the former is a step forward in terms of player evaluation. In his 2009 book *HockeyNomics*, Darcy Norman has a tremendous opening chapter that walks the reader through the basics of player evaluations using GC, among other statistics.[9]

Let's consider a specific example. In 2016–17, Connor McDavid won the Art Ross Trophy by having the most points in the regular season. As shown in the following table, he scored 30 goals (G) and 70 assists (A) for 100 points (PTS), which was 11 more than Sidney Crosby, who was tied for second place with 44 goals and 45 assists for 89 points.

However, the two players were less than a goal apart in terms of goals created (GC), since so many of McDavid's points were assists. We calculate McDavid's GC by multiplying his 30 goals by 0.5 to get 15 and then multiplying his 70 assists by 0.3 to get 21, for a total of 36. Actually, it works out to 35.9 instead of 36, because assists were technically worth 0.298 that season, but you get the picture.

NHL GOALS-CREATED LEADERS, 2016–17[10]

PLAYER	TEAM	GP	G	A	PTS	GC
Connor McDavid	EDM	82	30	70	100	35.9
Sidney Crosby	PIT	75	44	45	89	35.4
Nikita Kucherov	TBL	74	40	45	85	33.4
Patrick Kane	CHI	82	34	55	89	33.4
Brad Marchand	BOS	80	39	46	85	33.2
Mark Scheifele	WPG	79	32	50	82	30.9
Nicklas Backstrom	WSH	82	23	63	86	30.3
Vladimir Tarasenko	STL	82	39	36	75	30.2
Leon Draisaitl	EDM	82	29	48	77	28.8
Auston Matthews	TOR	82	40	29	69	28.7

I use a lot of compound statistics throughout this book, and I use this one to try to find the world's greatest women's hockey player. That chapter will also include a rather interesting way of taking opportunity into account and placing stats in context: GCAX.

9. Darcy Norman, "Why Bother with HockeyNomics?," *HockeyNomics* (Lone Pine Publishing, 2009), 41–110.

10. Raw scoring data for my calculations from NHL.com - Stats, National Hockey League, updated July 7, 2017, https://www.nhl.com.

GCAX is not an alien from *Star Trek*. Well, maybe it is, I'm not sure, but, in this case, it stands for goals created above expectations. In fact, that's often what the AX suffix means in hockey statistics. As you might have guessed, GCAX is a relative statistic. Just as the relative statistic that we built for the Chicago Blackhawks allowed us to compare their relative scoring rate of +0.50 goals per game to teams from different seasons, GCAX allows us to compare players from different teams.

In this case, GCAX is a player's goals created minus the average goals created of the other players at the same position on his or her team. Just like league scoring rates were very different for the Chicago Blackhawks in 2012–13 than they were for the Edmonton Oilers in 1983–84, players' scoring rates can vary greatly from one team to the next. (We'll flush out GCAX in more detail in our chapter on finding the world's best women's hockey player.)

Goaltenders

Goaltending analysis is actually the easiest to understand, since most of hockey's common statistics have already taken opportunity into account.

For example, goals-against average is the number of goals that were allowed per 60 minutes (the formula is goals allowed divided by minutes played multiplied by 60 minutes), which is exactly the type of statistic we explored with teams. And then there's save percentage, which takes opportunity into account by dividing a goalie's saves by the number of shots faced. That's exactly what we did with winning percentage and points percentage. As such, we can easily use these statistics to create relative stats.

The example we explored with Ovechkin and Stamkos also applies to goaltending statistics, like save percentage. That is, the same four options exist to deal with the factors that can affect save percentage, like how often a goalie plays short-handed.

1. You can ignore the impact of manpower and just compare goalies using overall save percentage. This works fine if they're on the same team or on teams that take roughly the same number of penalties.

2. You can break down a goalie's save percentage into short-handed save percentage, even-strength save percentage, and power-play save percentage. In fact, it's quite common to compare goalies using even-strength save percentage only.

3. You can combine the save percentages in each category to create a league-average ratio (which is called manpower-adjusted save percentage, or MASP). This makes perfect sense for goalies, since they usually play entire games, meaning that they all have ample data in all manpower situations. There are no goalies who are used exclusively in one manpower situation or another.

4. You can calculate the greater difficulty of saving shots when killing penalties in order to create an adjusted form of save percentage. In fact, there are forms of adjusted save percentage that also take into account the location of the shot and how many of them were rebounds or shots off the rush, which have a far greater chance of scoring.

If you're already familiar with save percentage (and the examples in this chapter mostly make sense), then you're already equipped to handle almost anything that will be thrown at you in the world of goalie stats (including a later chapter on finding the NHL's most valuable goalie).

Closing Thoughts

I may be pretty good with numbers, but I'm still familiar with math anxiety. I have experienced plenty of situations in which I got overwhelmed trying to wrap my head around some crazy formula or solve a problem that I could barely understand. Over time, I've learned that the solution to math anxiety is to break it down into smaller steps. The idea is to separate a larger problem into several components, solve the pieces, and then put everything back together. If the components are still too complicated, then just break them down further. Eventually, everything in math can be broken down to one plus one, which still equals two by the way.

While this chapter focused on providing the basics required to enjoy this particular book, these concepts apply to practically anything you're likely to encounter in hockey analytics. There are plenty of ideas that require a great deal of time and effort to break down and wrap our little walnuts around, but we'll always get there if we stay confident and have fun. I hope this overview will help you do just that.

WHO IS THE MOST
VALUABLE GOALIE?

With rare exceptions, no player has as significant an impact on a team's fortunes as the goalie, and yet it is outrageously difficult to pick the right one. While choosing the wrong player at another position probably won't affect the bottom line too dramatically, selecting the wrong goalie can easily derail a team's entire season.

Let's take an example. Setting aside the benefit of hindsight, flash back to the summer of 2016, and imagine that an organization could have had its pick of any goalie in the NHL. Who should they have chosen and why?

- Roberto Luongo of the Florida Panthers was one of the league's best goalies in 2015–16, but he was also the oldest. How much longer could he continue to perform as a solid number one goalie?
- Henrik Lundqvist of the New York Rangers was consistently among the top contenders for the Vezina, but he carried one of the highest cap hits and would for many years to come. Imagine getting stuck with his mediocre 2016–17 performance until 2020–21.
- Carey Price of the Montreal Canadiens would have been a great choice, and he was already established as an excellent goalie on a good contract. However, he carried a high injury

risk and had only two years remaining on his value contract. Now that we know he'll carry the greatest long-term cap hit among goalies, would he still be the right choice?

- The goalie who did wind up getting injured was Jonathan Quick of the Los Angeles Kings. Like Corey Crawford, he had had great playoff success and had won the Stanley Cup several times, but he hadn't really been among the best in the regular season. The Kings have now missed the playoffs in two of the past three seasons.

- Braden Holtby of the Washington Capitals would have been an excellent choice, but he hadn't established his credentials as clearly and reliably as the aforementioned veterans. Taking a chance on a goalie like that could just as easily have meant taking Cory Schneider of the New Jersey Devils instead, who had shown the same tremendous promise and had nearly identical stats as Holtby.

- Looking to the future, a team could have opted for a leading prospect, like Andrei Vasilevskiy of the Tampa Bay Lightning, Ilya Samsonov of the Washington Capitals, or Connor Hellebuyck of the Winnipeg Jets. This is the gambler's choice, since each of them has the potential to provide years of high contractual value—or possibly no value at all.

- Without the benefit of hindsight, few teams would have toyed with the idea of choosing Devan Dubnyk of the Minnesota Wild, Cam Talbot of the Edmonton Oilers, or Sergei Bobrovsky of the Columbus Blue Jackets. However, they were among the most valuable goalies at various points of the 2016–17 season, and they certainly appeared to be great choices at the season's conclusion.

What this demonstrates is that each goaltender has his own combination of strengths and weaknesses, which ebb and flow with each passing season. Determining who is the most valuable goalie in the league at any given time is an exercise in finding the right balance between each of these factors and projecting them over the long term. (Oh, and getting crazy lucky.)

We can't do anything about luck, but we can build a helpful model that identifies and weighs all of these factors and can identify the goalies who are most likely to be among the league's most valuable.

As always, the first step is to define our terms. In this case, what do we mean by valuable? Long-time readers know that value is defined a little bit differently in the world of hockey statistics than it is for the traditional MVP awards. After all, awarding Carey Price the Hart Trophy as the league's most valuable player in 2014–15 said just as much about Montreal as it did about Price himself.

For example, would Price have been as valuable with the New York Rangers, who have coped just fine without Henrik Lundqvist in peak form? The Rangers had the second-best record in the league in his absence during the 2014–15 season, 18-4-3, and were 21-10-2 with backup Antti Raanta in 2015–16 and 2016–17. In contrast, Montreal suffered an epic collapse in 2014–15 in Price's absence before resuming their position atop the Atlantic immediately upon his return.

Similarly, would an exceptional goalie like Price be the league's MVP in Los Angeles, whose exceptional team defence allows only 27 shots per game, most of them of low quality? After all, they did just fine with a replacement-level goalie like Peter Budaj in 2016–17. In contrast, goalies like Mike Condon, Ben Scrivens, and Dustin Tokarski were absolutely shelled in Montreal.

So when Price won the Hart as the league's most valuable player, it said something not just about Price but about Montreal. He probably wouldn't have won that award with the Rangers or the Kings. However, in this context, a goalie's personal value should ideally be measured in a way that's independent of the team for which he plays.

The other key aspect of a player's value is how much he is paid. There's a limit to how much teams can spend on their players, so if a goalie is paid more than he is worth, then that leaves less money to fill the rest of the roster. In other words, a team might be better overall with a weaker goalie, provided he costs less and the team uses that extra cap space to get a really good defenceman.

There may be a lot to consider when searching for the league's most valuable goalie, but fortunately we can begin from the basic, high-level approach of measuring a player's value that was crafted for skaters in the opening chapter of *Stat Shot*.[11] Briefly, the process is to establish a goalie's current performance level by capturing it in a single number, calculating a weighted average of his most recent seasons, figuring out how much random variation is in that result, and then regressing the result toward the league average to remove the impact of the random variation. That should smooth out the ups and downs of a goalie's career.

To project a goaltender's future performance, an age curve is applied to account for his natural development in his very early years, his gradual decline throughout his 20s, and then the more significant decline in his 30s. That should help account for the untapped potential of up-and-coming goalies while also capturing the risk of investing in older goalies.

Once we've established a career performance curve, we need to score it relative to the second component of a player's value: his cap hit. In essence, a goalie's performance level is measured relative to the value a team would receive if it invested the same cap space elsewhere. That should address those on lucrative contracts, like Lundqvist and Price.

In the end, a goalie's current value is derived as the simple sum of his cap-relative performance throughout the duration of his current contract. See, nothing could be simpler! Well, maybe not, but the entire process can be broken down into five clear and manageable tasks, each of which illuminates a particular aspect of the question.

In this chapter, we're going to flush out all of the above steps into a full-fledged projection model for goalies, including estimates of future games played and save percentage. This system will include accounting for random variation and age, will bring in goaltending data from the AHL, and will account for individual cap hits. In the end, the model itself will be more valuable than the actual answer to our question—and possibly just as interesting.

Remember that in the salary cap era, the league's most valuable goalie won't necessarily be its *best* goalie (a topic that was explored in

11. Vollman, "What's the Best Way to Build a Team?," *Stat Shot*, 12–74.

previous editions of *Hockey Abstract*[12]). The most valuable goalie will be the one signed to the longest and most reasonable contract and with the highest probability of sustaining strong, near-elite performance over the duration of that contract.

Let's start with the most important step: figuring out how effective each of the NHL's goalies are right now by predicting how they will perform in the next season.

Projecting the Next Season

Predicting how a goalie will perform in any given season involves examining the past, identifying all of the outside factors that would influence those results, and then filtering these influences out until only the goalie's own skill remains.

There are at least a dozen significant factors that can influence a goalie's save percentage.[13] Most notably there is our old nemesis random variation, but there's also special teams play, shot quality, recording bias, score effects, and various other team effects.

In this case, we're only going to deal with the two most important factors: random variation and special teams. Why? Because many of the following steps involve using historical data, and the information required to account for all those other factors wasn't available until recently. Once you build the basic system, however, it's easy enough to add in whatever additional factors are warranted (and feasible) in the future.

Quite frankly, tackling random variation is the biggest hurdle. Without question, the greatest obstacle to everything that has ever been explored in these pages is the fact that outcomes are the result of more than just the skill of the players and teams involved along with the outside factors we can measure. Some fans call it heart, others call them intangibles, and still more chalk it up to puck luck.

---

12. Rob Vollman, "Who Is the Best Goalie?" *Hockey Abstract* (CreateSpace Independent Publishing Platform, 2013), 43–51; Rob Vollman, "Who Is the Best Goalie?," *Hockey Abstract 2017* (CreateSpace Independent Publishing Platform, 2017), 134–163.

13. Rob Vollman, "Who Is the Best Puck-Stopper?" *Stat Shot*, 169–227.

Whatever you call it, these skewing effects are unpredictable and can only be sorted out after the fact.

As a graphical demonstration of the great obstacle that is random variation, consider the following chart of the rolling one-year save percentages of three of the top goalies from 2010–11 to 2016–17: Lundqvist, Price, and Boston's Tuukka Rask. These three goalies were each in their prime, didn't change teams, played in the Eastern Conference, and dominated both the save percentage leaderboard and the Vezina races. And yet, their individual save percentages bounced around like a Ping-Pong ball. That's due to random variation, and that's exactly why random variation really, really sucks.

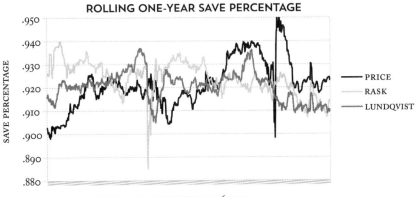

FROM 2010-11 THROUGH 2016-2017

Study this chart for as long as you want—which sections represent the *real* Lundqvist, Price, and Rask? Most fans will gravitate toward the most recent results on the far right (a tendency that is aptly named recency bias), but that's really no better a guess than any other arbitrary vertical slice. What happened in the past will happen again, but the problem is figuring out just what exactly did happen in the past.

There's a well-known method for using historical results to predict the future, and it has absolutely nothing to do with hockey. It was developed by 19th-century scientists like Auguste Bravais and Sir Francis Galton in entirely unrelated fields. In the sports world, this method has been used in baseball for decades.

The process boils down to identifying how much random variation is involved in historical results and removing it by regressing the numbers to the mean. The theories behind this process are quite complicated (and far outside the scope of a fun book like this), but the application is actually very easy.

How does it work? As explained in more (and characteristically excessive) detail in *Stat Shot*, every recorded observation of any kind whatsoever is a result of a repeatable skill and random variation.[14] For example, even when the environment is accounted for, save percentage is still a combination of the goalie's own abilities and puck luck. As such, the first step is to figure out just how much random variation is involved in each goalie's recent results. There are various ways of doing this, but they each involve one of those 19th-century concepts known as statistical correlation.

The most common way to figure out how much luck is involved in a set of results is to divide a goalie's data into two halves and see how closely correlated they are. At the risk of oversimplifying, the basic idea is that there is no random variation whatsoever if they have a perfect 1.0 correlation, and the observed results are entirely random if the two halves have a zero correlation. Most results obviously fall somewhere in between. This is called split-half correlation.

Another common way is to simulate a goalie's statistics using a randomized method and calculate how well it correlates to the actual results. If it's bang on, then the observed results are completely random. If they have no correlation, then the observed result has no random variation. Most results will fall somewhere in between.

Neither method requires a complicated formula, since most statistical packages—and even a simple Excel spreadsheet—will have a built-in correlation function that will do the calculations for us.

Let's pause for one moment though. In the quest to keep this simple and to the point, I'm glossing over some of the complexities, flying over some pretty important assumptions, and seriously irritating the more learned statisticians among you.

For example, I'm assuming that goaltender performance fits a normal distribution curve when it actually doesn't. It's much closer to a log-normal distribution. While I highly recommend reading up on statistical correlation and distribution curves, I concede that it's

---

14. Rob Vollman, "What's the Best Way to Build a Team?," *Stat Shot*, 12–74.

not everyone's idea of a great way to spend a Saturday night. Most of the time, the approach I'm describing will suit our purposes just fine.

And don't be thrown by the use of the word *random*, which can sometimes confuse the issue. It's easy to understand what *random* means when the discussion involves throwing dice or flipping coins, but it's not as clear when it involves professional athletes competing in a game of skill. To make matters worse, the word *random* is often misused as a synonym for assorted or miscellaneous. For example, you're not actually rolling dice when you go to the store for some random items, nor are you flipping coins when deciding to perform a random act of kindness, but we use that word anyway. Even hockey statistician Kent Wilson entitles his weekly column about specifically chosen topics "Random Thoughts" (although that's just to annoy me, he chortles).

To be clear, the statistical definition of *random* is being used here, "of or characterizing a process of selection in which each item of a set has an equal probability of being chosen."[15] In essence, some goals are scored as a direct consequence of the skills of the shooters, players, and goalies involved, while other goals are essentially selected through a random process of deflections and bounces off of skates, sticks, posts, and legs and the exact timing of blown whistles. It reminds me of the randomness of the Plinko game on *The Price Is Right*, which is precisely the type of random variation that we seek to identify—and remove—in order to reveal only the goaltender's true skill.

Once you've established the impact of random variation on the results using one of the aforementioned methods, you must remove it using statistical regression (another term that gets mentioned more and more every year).

Statistical regression is quite complex in theory but easy to implement in practice: the square root of the correlation that we previously calculated is the key number. That's the share of a goalie's performance that is skill-based, so we multiply that by the goalie's observed performance and then replace the rest with the league average.

In a sense, statistical regression is the process of removing the impact of random variation by "padding" a goalie's results with

---

15. Definition from http://www.dictionary.com/browse/randomly.

additional shots that we assume would have been stopped at a league-average rate. Two of those most responsible for teaching me how to apply these concepts to hockey, Tom Tango and my *Stat Shot* co-author Iain Fyffe, take it a step further and pad the results to a certain threshold. For example, Keith Kinkaid stopped 1,410 of 1,543 shots as New Jersey's backup from 2014–15 to 2016–17. Tango or Fyffe might pad that result to a threshold of 6,000 shots by adding 4,457 shots, of which a league-average goalie would have stopped 4,078 for a final result of 5,488 saves in 6,000 shots. Meanwhile, starter Cory Schneider saved 4,262 of 4,628 shots, which would get padded to a far lesser extent, to a total of 5,517 saves in 6,000 shots. From that, someone could infer that Schneider is worth 29 goals every 6,000 shots relative to Kinkaid.

The Tango-Fyffe approach has the advantage of being simple, and it also regresses large volumes of data far less than smaller volumes, or possibly not at all. Obviously, the more data you have, the less impact random variation will have on the observed results. That's why our trademark has always been to use three seasons' worth of individual player data rather than a single season.

To reduce the need to pad the results as described, we would ideally go back more than three seasons in order to include more data, but that would be at the risk of including less relevant data. To account for the decreased relevancy, the three seasons that we are using don't need to be weighted evenly. Since each year is generally twice as useful at predicting the future as the year before, a 4-2-1 weighting is commonly used for such purposes, as each digit has twice the value of the next. (That is, the 4-2-1 approach places more weight on recent data by multiplying the current year's data by four, the previous year's data by two, and the year before that by one.) However, some analysts weigh all three years equally, and Tango himself uses a 5-4-3 weighting for baseball's Marcel Method of projecting future performance.[16]

Eric Tulsky, now of the Carolina Hurricanes, built a model to figure out exactly how many years to use and how much weight to

---

16. Tom Tango, "The 2004 Marcels," *Tango Tiger* (blog), March 10, 2004, http://www.tangotiger.net /archives/stud0346.shtml.

place on each one.[17] His three-year weighting was 10-5-4, but he preferred a four-year model that was weighted 10-6-5-3. His work was reproduced by a blogger named garik16, whose talent for using historical data to project future performance is well-known within the hockey analytics community.[18] (The Tulsky-Garik method predicts the three following seasons, not just the next season, but the results, and the principle, remain the same.) Yet another blogger, going by the handle Fooled By Grittiness, calculated that 10-8-3-2 is the ideal weighting.[19]

Regardless of the exact weighting, the common thread is everybody basing their projections on three or four seasons rather than just one. To graphically demonstrate the value of doing so, consider the same chart as on page 39, but this time using a three-year rolling average. As you can see, this appears to be a far more accurate picture of each goalie's current performance level and the direction in which it is trending.

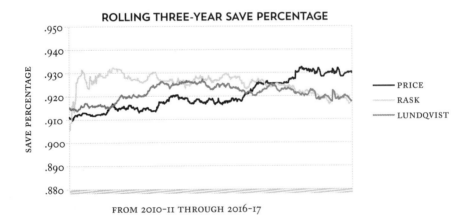

ROLLING THREE-YEAR SAVE PERCENTAGE

FROM 2010-11 THROUGH 2016-17

17. Eric Tulsky, "Quantifying the Added Importance of Recent Data," *Outnumbered* (blog), January 21, 2014, http://www.sbnation.com/nhl/2014/1/21/5329992/nhl-stats-projections-data.

18. garik16, "Projecting Future Goalie Performance: Updating and Improving Hockey Marcels," *Hockey Graphs* (blog), April 19, 2014, http://hockey-graphs.com/2014/04/19/projecting-future-goalie-performance-updating-and-improving-hockey-marcels/; garik16, "Forecasting Future Goalie Performance with Four Year Hockey Marcels," *Hockey Graphs*, February 3, 2014, http://hockey-graphs.com/2014/02/03/forecasting-future-goalie-performance-with-four-year-hockey-marcels/.

19. Fooled By Grittiness, "Marcel Projections for Goalies," *Fooled By Grittiness* (blog), September 14, 2016, https://fooledbygrittiness.blogspot.ca/2016/09/marcel-projections-for-goalies.html.

Since there is a remote but reasonable possibility that the most valuable goalie is someone without three consecutive NHL seasons, we need a way to pull in data from other leagues. This is explored in an upcoming chapter ("How Can We Compare a Player's Stats Between Leagues?"). For now, I'll just say that AHL data can definitely be used in this model, but only with some slight adjustments. Adding the AHL data can help us get a more accurate read on goalies like Jacob Markstrom of the Vancouver Canucks, Anton Forsberg of the Chicago Blackhawks, Philipp Grubauer of the Washington Capitals, Matt Murray of the Pittsburgh Penguins, Mike McKenna of the Dallas Stars, and Michael Leighton and Peter Budaj of the Tampa Bay Lightning.

However, the lack of data still results in low confidence for anyone coming directly from another league without much time in the AHL or NHL, like ECHL goalies Chris Driedger of the Ottawa Senators, Scott Darling of the Carolina Hurricanes, and Mackenzie Skapski of the New York Rangers. There's obviously not much we can do about the lack of data for backup goalies who sit on the bench all year, like free agents Jonas Gustavsson and Al Montoya of the Montreal Canadiens. Fortunately, neither of those groups of goalies is likely to include the NHL's most valuable goalie, so we don't need to dwell on this limitation for the purposes of answering this chapter's question.

All of this work brings us to the first plateau, which is a reasonably accurate estimate of each goalie's true talent in the present. To summarize, we took a three-year weighted average of each goalie's save percentage and regressed it to the league average to remove the calculated impact of random variation. As it stands, Bobrovsky leads the pack, but not by much.

TOP-10 WEIGHTED AND REGRESSED NHL AND AHL SAVE PERCENTAGES, AFTER 2016–17

GOALIE	TEAM	SV%
Sergei Bobrovsky	CBJ	0.9239
Matt Murray	PIT	0.9237
Braden Holtby	WSH	0.9237
Carey Price	MTL	0.9236

GOALIE	TEAM	SV%
Devan Dubnyk	MIN	0.9226
Corey Crawford	CHI	0.9207
Craig Anderson	OTT	0.9206
John Gibson	ANA	0.9202
Cam Talbot	EDM	0.9189
Anton Forsberg	CHI	0.9189

It may seem unusual that Price is ranked fourth, but it's only by 0.0003, and it's primarily because of the uncertainty introduced from his having played only 12 games in 2015–16. With different weightings and regression factors, the top five could conceivably be presented in virtually any order.

That should capture how well goalies are playing right now, but how will their play change in the future, say over the course of their entire contract and/or entire career? Theoretically, goaltenders improve as they gain more experience and then decline as they age, but when and to what extent? We'll look at that next.

Adjusting for Age

What impact does age have on a goaltender's performance? Goalies like Roberto Luongo and Martin Brodeur appeared to be completely ageless, while others fell apart years before their 30th birthday. It may be impossible to predict how a single individual's performance will change with age, but we can statistically base a projection on the average historical results at each age.

When building age curves, the most common mistake is to overlook survivor bias. On a first attempt, most people build an age curve by grouping all goaltenders together by age and calculating each group's average save percentage. That may sound like a reasonable approach, but it absolutely won't work. It will produce an age curve that doesn't seem to indicate any change at all as goalies age.

Why? As Tulsky once explained, older goalies are phased out of the NHL as they decline and are replaced with younger, better goalies.[20] That means that the only goalies who will remain in the analysis (at any age) are those who are good enough to keep their jobs. In other words, the survivors. As such, this type of aging curve will include all kinds of goalies in their 20s but only the truly exceptional goalies in their 30s, and so both groups of goalies will appear to be roughly the same.

As explained in (excessive) detail in *Stat Shot*, the right way to build an age curve is to do it the same way they do in baseball, which is to calculate the average change in a goalie's performance from one age to the next.[21]

The following age curve was built by gathering all 1,035 goalies who played at least 20 games in any two consecutive seasons between 1983–84, when save percentage was first recorded, and 2014–15. (As of September 2017, NHL save percentage data goes back to the 1954–55 season.) The average change in save percentages as each group of goalies went from one age to the next, relative to the league average, is marked with a dot. For example, the group of 30-year-old goalies saw their save percentages drop by just under 0.001, while the save percentages of the 34-year-old goalies dropped by over 0.005.

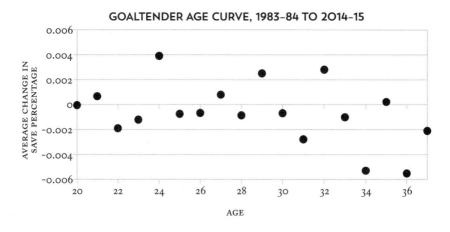

GOALTENDER AGE CURVE, 1983–84 TO 2014–15

20. Eric Tulsky, "How Goalies Age, and the Survivorship Bias Problem," *NHL Numbers* (blog), December 6, 2012, http://nhlnumbers.com/2012/12/6/goalie-aging-survivorship-bias.

21. Rob Vollman, "What's the Best Way to Build a Team?," *Stat Shot* (2016), 12–74.

Even with the largest possible sample size, the results are unfortunately not very smooth and clean. However, it's still perceptible that goalies tend to collapse in their mid-30s and that, until then, age doesn't affect a goalie's save percentage by more than a single point (i.e., 0.001) per year. Specifically, the average improvement of a goalie throughout his 20s is a negligible 0.00035 per season, it drops by an almost inconsequential 0.00063 per season in the early 30s, but then it collapses by 0.0031 starting at around age 34 until the goalie's career is over.

There's no consensus on this point in the hockey analytics community, mind you. Some statisticians have calculated very similar, modest results, like Fooled By Grittiness, while others have identified a slightly more noticeable impact due to age, like Steve Burtch and garik16.[22] The only common thread in each set of results is the collapse that occurs in the mid-30s, which usually signals the end of a goalie's career.

When Do Careers End?

Once we've established a goalie's current performance level and adjusted for age, the next step is to determine if his career is already over. If it is, we've finished our projection. If his career isn't over, we continue our projection into the next season.

This determination isn't as simple as waiting until a goalie's value drops below the proverbial zero. First of all, it's not that easy to figure out where "zero" really is. Secondly, goalies are subject to peaks and valleys, and it can take time to figure out if a goalie's performance has actually fallen below that arbitrary point or if he should have another season to redeem himself. Finally, the great goalies never truly reach zero and, ultimately, retire at a chosen age due to wear and tear or simply personal preference.

---

22. Fooled By Grittiness, "Marcel Projections for Goalies," *Fooled By Grittiness* (blog), September 14, 2016, https://fooledbygrittiness.blogspot.ca/2016/09/marcel-projections-for-goalies.html; Steve Burtch, "Theoretical Goaltending Aging Curve," *Pension Plan Puppets* (blog), September 5, 2013, https://www.pensionplanpuppets.com/2013/9/5/4696042/theoretical-goaltending-aging-curve-Bernier-Reimer; garik16, "How Well Do Goalies Age? A Look at a Goalie Aging Curve," Hockey Graphs, March 21, 2014, https://hockey-graphs.com/2014/03/21/how-well-do-goalies-age-a-look-at-a-goalie-aging-curve/.

To shed some light on when goalies tend to pack it in, I grabbed the final NHL season for every goalie since 1983–84 and organized the data into the following chart. The two key factors in determining retirement are age and performance. Age is straightforward, and performance is measured using save percentage. Since the league-average save percentage climbed steadily until the year 2000 and has occasionally spiked (1993–94) and dipped (2005–06), we calculate each goalie's save percentage relative to the league average that year. For example, a value of −0.010 on this chart refers to 0.010 below the league-average that particular season. For the 2016–17 season, when the league average save percentage was 0.913, that would mean the goalie's save percentage was 0.903.

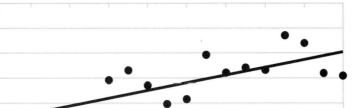

FINAL SEASON PERFORMANCE BY AGE

As the trend line demonstrates, goalies who are pushing 40 need only drop 10 points below league average, or 0.903 in the modern day, for their careers to come to an end. Younger goalies have to be a dismal 0.893 for their careers to end. That makes intuitive sense, as a 27-year-old isn't likely to retire after one bad season, but a 35-year-old who dips even a little bit below average is likely to get passed over for a younger option.

In our model, when a goalie's projected performance drops below the point on the 0.903 line that corresponds to his age, we can assume he will retire and his career projection is complete.

While these results are good enough for our model, it's important to remember that every individual case is different. For example,

Jacques Plante played in the NHL until age 44, and after that he continued in the WHA until age 46. More recently, Dominik Hasek was an elite NHL goalie until age 42, after which he continued to compete in the European leagues until age 46.

On the flip side, the wear and tear of NHL action, combined with each goalie's personal preferences, can lead to retirement long before the Plante/Hasek years. Patrick Roy, for instance, retired at age 37, despite finishing fourth in Vezina Trophy voting. That was purely a personal decision and not one that any model could predict. When Martin Brodeur was that same age, he finished third in the Vezina voting and continued to play five more seasons—all but one of which were below that 0.903 line, which normally triggers retirement. His reputation allowed him to continue playing almost indefinitely.

Based on their age and performance level in 2014–15, it made perfect sense for goalies like Ilya Bryzgalov, Evgeni Nabokov, and Martin Brodeur to (finally) retire. Ray Emery, Viktor Fasth, and Niklas Backstrom didn't officially retire, but they haven't seen any NHL action either. Al Montoya, Ben Scrivens, and Chad Johnson were the only goalies beyond the retirement line who kept playing, which was obviously the right decision in Montoya's and Johnson's case; Scrivens left the NHL after the following season. In 2016–17, goalies like Antti Niemi, Michal Neuvirth, and Justin Peters should have considered retiring, and by the time you read this book, you'll know whether the right decision was made.

The remaining goalies above the retirement line are assumed to be playing another season, so we can move on to the final step, which is to calculate their value by comparing their expected performance to their cap hit. After all, value is determined not just by how good a player is but how good he is relative to the cap space he is using.

Converting to Goals and Dollars

On the surface, converting a goalie's save percentage into the number of goals he prevented seems rather easy. After all, every additional save past a certain threshold equals one goal that was prevented, right?

-CONFLICT IN THE SITUATION ROOM-

It's actually not quite that easy, since there are two key questions. First, how many shots is a goalie going to face and, therefore, how many saves will he make? Second, what threshold should we compare his performance against?

Without diving too deep into these two rabbit holes, we can crudely estimate games played and shots faced by using the same weighted average of a goalie's three previous seasons that we used for estimating save percentage, provided that it actually adds up at the team level for all known contracts. Ideally, the model would also include similar adjustments for age and performance, which would increase a goalie's workload as he improves and matures and decrease it as he ages and declines.

As for the threshold against which to compare each goalie, there are a few ways to go about this. Most analysts start by comparing a goalie's performance to the league-average save percentage and multiplying that by shots faced to arrive at the number of goals prevented relative to the league average. In fact, there's a statistic called goals saved above average, or GSAA.

For example, Lundqvist had an expected save percentage of 0.9183 for the 2015–16 season. When multiplied by the expected 1,567 shots, he was expected to make 1,439 saves. Subtract the number of saves by a goalie with a league-average save percentage of 0.9136, which is 1,432, and Lundqvist was expected to have prevented 7 more goals than an average goalie. In reality, Lundqvist faced a league-high

1,944 shots and had a 0.9198 save percentage, so his 1,788 saves were 10 more than a league-average goalie.

Saving an extra 10 goals sounds great, but remember that Lundqvist has a cap hit of $8.5 million, which was the highest in the NHL among goalies at the time, and far higher than the league average of roughly $3.2 million. That means that each additional goal he prevented cost the Rangers around $530,000. Is that good value, or would the Rangers have been better off with a league-average goalie who would have allowed 10 more goals but who would have spared them an extra $5.3 million to spend elsewhere in the roster, either on players who would have generated more scoring or prevented more shots? After all, that's what value, and being valuable, is really about.

Rather than compare a goalie's numbers to the league average, many analysts prefer to compare them to the level of goaltending that is freely available to replace him if he were injured. For example, when Jonathan Quick was injured, there were no league-average goalies available. Instead, Peter Budaj was called up from the AHL to replace him, which makes goalies like him more useful and/or appropriate as a comparison point to establish Quick's true value.

Using the league average as the comparison point or the Budaj-replacement level each have their advantages and disadvantages. While the performance level of a league-average goalie is known, the level of freely available call-ups is not. Since there are so many different ways to identify and measure so-called replacement-level goaltending, many of which were listed and explored in *Stat Shot*, there's still no consensus within the hockey analytics community of exactly where this threshold actually is and how to calculate it.[23]

However, the problem with comparing goalies to the league average becomes clear when it's time to convert goals into dollars. The cost of a replacement-level goaltender is known; it is the league minimum salary for any given season (or, perhaps, very slightly above it), but the cost of a league-average goalie isn't as obvious. It is not simply the league-average salary, since salaries are not distributed evenly and particularly good or bad contracts can skew the numbers.

So pick your poison. Whichever approach you adopt, there will

23. Rob Vollman, "Who Is the Best Puck-Stopper?," *Stat Shot*, 169–227.

be some slight uncertainty surrounding either the performance level or the cost. In our case, we calculated the following results relative to replacement level, so they may therefore suffer slightly from the aforementioned uncertainty about exactly where that is.

Results

And here we are at last. After estimating every goalie's true talent level by taking a weighted average of the past three seasons of NHL and AHL data, regressing it back to the league average based on the number of shots faced, then extrapolating each goalie's future based on how other goalies of that age have fared historically, and comparing the goals saved to what is expected from a goalie with the same cap hit, we finally arrive at a list of the 10 most valuable goalie contracts, as of the end of the 2016–17 season.

TOP-10 MOST VALUABLE NHL GOALIES, AFTER 2016-17

AGE	GOALIE	TEAM	CAP HIT	EXPIRY	VALUE
31	Devan Dubnyk	MIN	$4.33M	2021	68.7
28	Braden Holtby	WSH	$6.1M	2020	48.0
30	Cam Talbot	EDM	$4.17M	2019	23.5
24	John Gibson	ANA	$2.3M	2019	22.8
33	Corey Crawford	CHI	$6.0M	2020	21.4
28	Frederik Andersen	TOR	$5.0M	2021	19.4
23	Matt Murray	PIT	$3.75M	2020	18.0
29	Sergei Bobrovsky	CBJ	$6.0M	2019	17.3
29	James Reimer	FLA	$3.4M	2021	15.2
30	Carey Price	MTL	$6.5M	2018	8.5

From this perspective, Dubnyk and Holtby are the two most valuable goalies in the NHL right now. Neither choice should generate much controversy from either within or outside the stats community.

Different models will yield slightly different results. Playing with how previous seasons are weighted or how much the values are regressed toward league average or what a goalie's workload will be or how good replacement level is for goalies may change the order of the next six or seven positions on the list, but every model should have Holtby or Dubnyk on top, followed by the same six or seven names.

At first glance, it may seem odd to find Price in 10th. He may be (much) more valuable on a per-season basis, but he is only signed for one more season at this level. After this coming season, his cap hit increases to an incredible $10.5 million, which will wash away most of his value. As such, he's not as valuable as lesser goalies who have better and longer contracts, especially those who are younger, like John Gibson or Matt Murray or even Antti Raanta of the Arizona Coyotes or Andrei Vasilevskiy of the Tampa Bay Lightning, who were just barely beaten out by Price on this list.

And no, Henrik Lundqvist is not in the top 10. In fact, he may be the league's least valuable goalie, especially after his numbers were pulled down by a surprisingly mediocre 2016–17 season. He's 35 and signed to a cap hit of $8.5 million for four more seasons, at which point he may start to cost the team as many goals as he has ever saved them. He's a great goalie, but not a valuable goalie from this perspective.

Closing Thoughts

Models like these are no trade secret. Most teams with even a minimal investment in statistical analysis can easily incorporate at least some of these strategies into their own player and contract evaluations.

Since most organizations can build models like these, the competitive advantage is in the details and how each of the individual pieces is executed. For example, there's no consensus on how many seasons to use in a goalie's projection and how to weigh them, nor is there a standard aging curve. The teams that get there first will have an advantage over the others.

There are many other opportunities to improve this model, such as incorporating data beyond the AHL and NHL, adjusting each goalie's save percentages for quality-related factors, much as Fooled By Grittiness has already explored, and accurately projecting when and how much a goalie's workload will increase and how that will affect his performance.[24]

For example, goalies like Arizona's Antti Raanta and Carolina's Scott Darling could double their respective single-season career highs of 26 and 27 starts this season. If they can retain their performance with that greater workload, then they could easily climb into that top-10 list.

But with the information we had in hand and the choices that we made, there's a strong case that either Dubnyk or Holtby is the league's most valuable goalie, and we can construct that compelling case for about a half-dozen others. Even with the potential improvements to the model, it's unlikely that the end result would change too dramatically. By the time this book hits the shelves, an extra season's worth of data will hopefully have made the answer more obvious.

---

24. Fooled By Grittiness, "Marcel Projections for Goalies," *Fooled By Grittiness* (blog), September 14, 2016, https://fooledbygrittiness.blogspot.ca/2016/09/marcel-projections-for-goalies.html.

HOW CAN WE COMPARE A PLAYER'S STATS BETWEEN LEAGUES?

Sample size is critical to the effective statistical evaluation of hockey players, and one way to expand it is to find ways to look at data beyond the NHL.

As a rule of thumb, a player's individual performance should be based on three full seasons. Anything less, and the results are going to be as much a consequence of random variation and outside factors as the player's own skill. Anything more risks including less relevant data. The problem is that over half the players involved in any given NHL game will have played fewer than 200 games in the league.

That's where NHL league translations can help. To calculate a translation, we take a player's scoring totals for the target league and multiply them by the appropriate factor from the table on the next page. This will give you a rough equivalent of what that player would have achieved in the NHL. You can combine the resulting numbers, referred to as the player's NHLe (NHL equivalent), with any existing NHL data to help extend your understanding of a player's current performance level to at least three seasons, which you can then use as part of a statistical projection.

NHL TRANSLATION FACTORS, AS OF 2016-17

LEAGUE	FACTOR
Kontinental Hockey League (KHL)	0.77
Swedish Hockey League (SHL)	0.62
American Hockey League (AHL)	0.47
Finland SM-liiga	0.46
Western Collegiate Hockey Association (WCHA, pre-2013)	0.44
National Collegiate Hockey Association (NCHC)	0.43
Switzerland (NLA)	0.43
Hockey East	0.38
Big 10	0.33
Central Collegiate Hockey Association (CCHA, now defunct)	0.32
Ontario Hockey League (OHL)	0.31
Western Hockey League (WHL)	0.28
Quebec Major Junior Hockey League (QMJHL)	0.25
ECAC	0.23

Let's consider a quick but meaningful example. Artemi Panarin scored 62 points in 54 games for St. Petersburg in the 2014–15 season. Multiply that by the KHL's translation factor of 0.8 (at the time), and he was expected to score 48 points in 54 NHL games or 71 points in 80 games.

Panarin scored 77 points in 80 games with Chicago in 2015–16, which is six points higher than his NHLe. Translations are not projections, but Panarin's incredible rookie scoring success wasn't completely unpredictable.

While not always within 8% as it was with Panarin, this simple approach is enough to get a pretty decent picture of how the majority of players from other leagues will perform in the NHL, whether they're rookies or veterans.

Of course, this isn't exactly the book for fans who are happy with 92% accuracy or satisfied with a "decent" picture. The following pages explore everything I've learned from working closely with NHL translations over the past 12 years, including how to refine the process for even greater accuracy, as well as some of the tips and tricks that I've picked up along the way.

Probably the most important development is learning how to account for age when building translations. Even a matter of a few months can make a big difference when dealing with players under age 20, and it's important to take those with so-called late birthdays into account. When a prospect significantly outperforms the translation in his rookie NHL season, it can often signal a big drop in year two and beyond.

We also know a great deal about the relative quality of the various leagues. For instance, we have learned that Russia is on the decline and that Finland is better than we had previously thought. We can also now separate league scoring levels from the quality levels of each league, and we have broken up leagues like the NCAA into their various conferences for greater precision.

Since every league is different in terms of the quality of its players, the coaching, the rules, and the size of the ice surface (among many other factors), we have learned a lot from a closer, individual examination of each league. We have also built up a list of the various types of players to use for comparison purposes, and we've learned a lot from studying some particularly noteworthy examples.

As for the basic translation process itself, basing projections on something other than a straight slope anchored at zero points, considering each player's ice time, and separating a player's scoring based on manpower situation are all recent improvements to the accuracy of these translations. Finally, we have crafted a goaltending translation system for the very first time, which we used in the previous chapter to project goalie statistics.

Even if you're already quite familiar with NHL translations from the original *Hockey Abstract* or elsewhere, there is still plenty of new developments to catch up on. However, we'll be starting by getting up to speed on what league translations actually are, where they come from, and how they work.

League translations is one of my favourite topics, dating all the way back to when Bill James introduced them to baseball in 1985 and Gabriel Desjardins to hockey in 2004.[25] With the possible exception of the historical player projections that use them, I can't think of any single, specific area of study to which I have devoted more time and energy.

That's why the most detailed chapter in the first *Hockey Abstract* was dedicated to this topic.[26] My passion was clearly contagious, and I was honoured when Tom Tango, the first fellow hobbyist that I ever met who had consulted with NHL front offices, referred to this chapter as "the most hard-core stuff" and that "it's an excellent presentation, and the best part of the book for me."[27] He, and all the other fans who have really embraced NHLe since then, have inspired me to revisit this topic and dig even deeper.

As explained in more detail in the aforementioned book, the story essentially begins with Bill James and his 1985 *Baseball Abstract*, where he introduced what he felt was his most important concept: major league translations. James found that his estimates of what a minor-league player's statistics would translate to in the major leagues had roughly the same predictive power for the following season as an equivalent set of actual major league data. In the case of batting average, for instance, it would fall within 25 points of the translation.

About 20 years later, Desjardins applied the same approach to hockey, by dividing the points per game of new NHL players by their points per game from the previous season, creating the original set of league translation factors. Even though they have fallen considerably out of date by now, many analysts continue to use those original translation factors or the ones Desjardins updated in 2006.

---

25. Bill James, *The Bill James Baseball Abstract, 1985* (Ballantine Books), 1985; Gabriel Desjardins, "League Equivalencies," Hockey Analytics, 2004, http://hockeyanalytics.com/Research_files/League_Equivalencies.pdf.

26. Rob Vollman, "Translating Data from Other Leagues," *Hockey Abstract*, 2013, 159–182.

27. Tom Tango, "Hockey Abstract," *Tango Tiger* (blog), September 8, 2013, https://tangotiger.com/index.php/site/comments/hockey-abstract.

Over a decade has passed, but Desjardins's original approach remains the backbone of today's method. The basic process to calculate the updated translation factors shown at the top of this chapter has five steps:

1. Find all of the players who went from the target league to the NHL within the selected time period and gather their scoring data.

This selected time period can be all of that league's history or, for the purposes of this chapter, since the 2004–05 lockout. Choosing longer periods results in a much larger sample of data, but it also risks including less relevant data, if the quality of the league has changed significantly over time.

Be sure to consider the direction in which the player was headed, since the relevant data points are commonly those players coming to the NHL, not those who are leaving it. This does introduce selection bias, but it is the right kind of bias, assuming the results are being used to evaluate players who might go to the NHL in the future.

While this chapter focuses entirely on scoring data, there's no reason why this same process couldn't be applied to other statistics, such as shots, penalty minutes, or catch-all statistics.

2. Normalize every player's data to a common scoring-level standard.

In this context, normalizing just means converting all of the numbers to a common standard. For example, scoring a lot of points in the high-scoring 1980s is a lot different from achieving the same scoring totals in the so-called dead puck era of the late 1990s and early 2000s.

Statistically, we normalize the data by dividing each player's points by the league's average goals per game that season, and then by multiplying it by the current league-average goals per game.

While often ignored in the name of simplicity, including in some of the tables that follow and even by Desjardins himself, this step is a key refinement because it helps distinguish between scoring changes that were caused by fluctuations in league scoring levels and those caused by changing leagues.

3. Calculate every player's (normalized) points per game for both leagues.

The simplest and most common method to take opportunity into account uses games played, but this chapter will also explore a more refined method that uses ice time.

4. Divide each player's NHL points per game by his points per game in the other league.

Typically, we only include players who played a certain number of games in both leagues in the calculation. The lower that threshold, the more random noise we risk introducing, but the larger our data set can be. On the other hand, the higher the threshold, the more selection bias we introduce, since only the most successful players will have played enough games to qualify. In this chapter, I chose a minimum of 40 games for the AHL and 20 games for the leagues with smaller sample sizes.

5. To arrive at the league translation factor, calculate the average of all the player translation factors from step four.

And that's it. The end result is a crude but effective way of estimating how much of his scoring a player will retain when he moves to the NHL.

Obviously, the most common use for this system is as part of a statistical projection for the league's rookies, so that's where we'll begin.

Prospects

Put away the tarot cards and tea leaves—league translations help form the basis of rookie scoring projections.

The big difference between calculating and using translation factors for prospects, rather than veterans, is accounting for age. As Iain explored in his development of the Projectinator in *Stat Shot*, it is far easier for the older players to score in the Canadian major junior leagues than it is for the younger teenagers. They're older, bigger, and stronger, and they get a lot more opportunity to play. He concluded that "recording a season of 100 points may be impressive for a 17-year-old, but a 20-year-old player doing the same has done nothing terribly special."[28]

To crudely illustrate this reality, consider the average translation factor for the Ontario Hockey League (OHL) broken down by age. For a variety of reasons, a 17-year-old is expected to retain over 40% of his scoring if he makes it to the NHL the following season, which is almost double what can be expected of a 20-year-old.

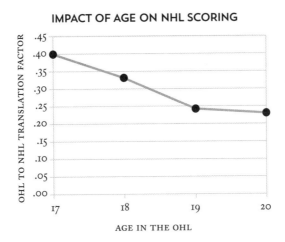

IMPACT OF AGE ON NHL SCORING

The effect isn't nearly as dramatic in the WHL and the QMJHL, but it still exists.

As for U.S. college hockey, the impact of age is not terribly significant beyond the first season (at age 18), because the great majority of players are already 20 or older. However, do keep your eyes peeled for players who dominate the game at age 18, like Jonathan Toews, Jaden Schwartz, and Dylan Larkin, because they tend to immediately score a lot of points in the NHL (unless they're 5-foot-8).

---

28. Iain Fyffe, "What Do a Player's Junior Numbers Tell Us?," *Stat Shot*, 75–118.

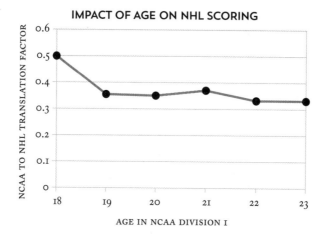

IMPACT OF AGE ON NHL SCORING

There are three basic ways to account for age when translating a player's scoring data. The obvious approach is to use separate translation factors for ages 17, 18, 19, and 20 and up (or 19 and up, arguably). However, for most leagues there's not enough data to even calculate an overall translation factor, let alone one that is broken down by age.

Another solution is to apply the same adjustments Iain uses in the Projectinator to normalize every player's scoring to a given age, such as that of an 18-year-old, and then calculate and/or apply the league's translation factor from there. (The Projectinator normalizes every player to a 17-year-old, which we've adjusted to 18 years old for our purposes here, since the translation factor has a stronger basis for that age.)

POINT MULTIPLIERS BY AGE, IN YEARS AND MONTHS[29]

	O	1	2	3	4	5	6	7	8	9	10	11
15	1.710	1.690	1.671	1.654	1.636	1.618	1.600	1.583	1.567	1.550	1.533	1.518
16	1.502	1.487	1.471	1.456	1.442	1.426	1.412	1.399	1.385	1.370	1.357	1.344
17	1.190	1.185	1.179	1.174	1.168	1.162	1.156	1.151	1.145	1.140	1.135	1.130
18	1.000	0.990	0.981	0.973	0.963	0.955	0.946	0.938	0.930	0.921	0.913	0.905
19	0.802	0.792	0.782	0.771	0.762	0.752	0.743	0.733	0.724	0.714	0.706	0.696
20	0.619	0.610	0.599	0.589	0.580	0.570	0.561	0.552	0.543	0.535	0.526	0.518

Finally, there's the option to study each league individually and to make a case-by-case translation based on past players with

---

29. Iain Fyffe, "What Do a Player's Junior Numbers Tell Us?," *Stat Shot*, 89.

comparable statistics at the same age. That's why each of the following sections for the Canadian Hockey League (CHL) is broken down by age and why age is included as the first column in the sections covering NCAA Division I hockey.

Before we begin, be aware of players with late birthdays. Since the cut-off used in the NHL is a player's age on February 1, while the age cut-off for the NHL Entry Draft is September 15, anyone born between those two dates is said to have a late birthday. Take careful note of such players, since they were drafted a year later than those born earlier in the same calendar year, which means they had more time to develop.

Ontario Hockey League

The largest of the CHL's leagues was formed in 1974 from the Ontario Hockey Association and has provided about a fifth of all NHL hockey players over the years.

However, most OHL players compete in a semi-professional league like the AHL before moving to the NHL. Right now, there are about half a dozen NHLers every season who played in the OHL the previous year, which has been the case since the early 1980s, before which the rate was two or three times greater.

The OHL's translation factor has gone through peaks and valleys over the years, and it's currently near its all-time high of 0.31.

Initially, it remained close to that high throughout the 1970s and 1980s, but it fell to nearly 0.2 through the 1990s because the so-called dead puck era affected the NHL far more than the juniors. This would be an excellent example of where normalizing the data before calculating the translation factor makes a lot of sense.

The OHL's translation factor rose to 0.23 in the early 2000s, and then dipped to 0.21 coming out of the 2004–05 NHL lockout, but it has steadily increased toward its high since then.

As demonstrated in the preceding section, age is critical when studying CHL players. The translation factor for 17-year-olds who competed in the OHL, were drafted, and then immediately moved up to the NHL at age 18 is a much higher 0.41. Why? Because of selection bias. Only the truly exceptional players will get the opportunity to play in the NHL at that young of an age, and only for teams who are so desperate for talent that an 18-year-old rookie is given top-six minutes. The eight players who have done so since the 2005–06 season are all excellent examples of such situations. (In all tables, the season and the first set of numbers refer to the player's totals in the target league, and the second set of numbers to the following season in the NHL.)

PLAYERS AGED 17 WHO WENT FROM THE OHL TO THE NHL, 2005-06 TO 2016-17[30]

PLAYER	SEASON	GP	G	A	PTS	TEAM	GP	G	A	PTS
Jeff Skinner	2009–10	64	50	40	90	CAR	82	31	32	63
Sam Gagner	2006–07	53	35	83	118	EDM	79	13	36	49
Steven Stamkos	2007–08	61	58	47	105	TBL	79	23	23	46
Jordan Staal	2005–06	68	28	40	68	PIT	81	29	13	42
Aaron Ekblad	2013–14	58	23	30	53	FLA	81	12	27	39
Ryan O'Reilly	2008–09	68	16	50	66	COL	81	8	18	26
Jakob Chychrun	2015–16	62	11	38	49	ARI	68	7	13	20
Zach Bogosian	2007–08	60	11	50	61	ATL	47	9	10	19

Minimum 20 games played in both leagues.

This isn't to argue that a translation factor of 0.41 should be used for *all* 17-year-olds. Just because a player is 17, it doesn't mean that he will retain as much of his scoring as Calder Trophy winners like Jeff Skinner or future franchise players like Steven Stamkos. It is far more likely that he will not play in the NHL for at least one more season.

Take Dylan Strome, for instance. He was drafted by the Arizona Coyotes third overall in 2015 after scoring 129 points in 68 games for the Erie Otters. With a 0.41 translation factor, that works out to 64 points in the NHL as an 18-year-old rookie. The fact that the Coyotes chose to keep him with Erie for two more seasons is

---

30. Unless acknowledged otherwise, all raw data in the tables in this section is from the National Hockey League, accessed May 23, 2017, https://www.nhl.com.

an obvious clue that the 0.41 translation factor only applies to very special players.

The next group of players is a mix of those who played one more season in the OHL after being drafted at age 18 and those who had late birthdays, got drafted, and turned 19 shortly after going directly to the NHL. It's pretty easy to spot who falls into the latter category, since they're almost all at the top of the list (and marked in italics).

PLAYERS AGED 18 WHO WENT FROM THE
OHL TO THE NHL, 2005-06 TO 2016-17

PLAYER	SEASON	GP	G	A	PTS	TEAM	GP	G	A	PTS
Patrick Kane	2006–07	58	62	83	145	CHI	82	21	51	72
Mitchell Marner	2015–16	57	39	77	116	TOR	77	19	42	61
Matt Duchene	2008–09	57	31	48	79	COL	81	24	31	55
John Tavares	2008–09	56	58	46	104	NYI	82	24	30	54
Gabriel Landeskog	2010–11	53	36	30	66	COL	82	22	30	52
Connor McDavid	2014–15	47	44	76	120	EDM	45	16	32	48
Matthew Tkachuk	2015–16	57	30	77	107	CGY	76	13	35	48
Taylor Hall	2009–10	57	40	66	106	EDM	65	22	20	42
Cam Fowler	2009–10	55	8	47	55	ANA	76	10	30	40
Michael Del Zotto	2008–09	62	13	50	63	NYR	80	9	28	37
Sean Monahan	2012–13	58	31	47	78	CGY	75	22	12	34
Nail Yakupov	2011–12	42	31	38	69	EDM	48	17	14	31
Alex Galchenyuk	2012–13	33	27	34	61	MTL	65	13	18	31
Olli Maatta	2012–13	57	8	30	38	PIT	78	9	20	29
Travis Konecny	2015–16	31	23	33	56	PHI	70	11	17	28
Mikkel Boedker	2007–08	62	29	44	73	PHX	78	11	17	28
Drew Doughty	2007–08	58	13	37	50	LAK	81	6	21	27
Bo Horvat	2013–14	54	30	44	74	VAN	68	13	12	25
Josh Bailey	2007–08	67	29	67	96	NYI	68	7	18	25
André Burakovsky	2013–14	57	41	46	87	WSH	53	9	13	22
Tyler Seguin	2009–10	63	48	58	106	BOS	74	11	11	22
Alexander Burmistrov	2009–10	62	22	43	65	ATL	74	6	14	20
Jared McCann	2014–15	56	34	47	81	VAN	69	9	9	18
Dougie Hamilton	2011–12	50	17	55	72	BOS	42	5	11	16
Jiri Tlusty	2006–07	37	13	21	34	TOR	58	10	6	16

PLAYER	SEASON	GP	G	A	PTS	TEAM	GP	G	A	PTS
Nikita Zadorov	2013–14	36	11	19	30	CBJ	60	3	12	15
Devante Smith-Pelly	2010–11	67	36	30	66	ANA	49	7	6	13
Lawson Crouse	2015–16	49	23	39	62	ARI	72	5	7	12
Tom Wilson	2012–13	48	23	35	58	WSH	82	3	7	10
Connor Carrick	2012–13	68	12	32	44	WSH	34	1	5	6
J.T. Miller	2011–12	64	25	37	62	NYR	26	2	2	4

Minimum 20 games played in both leagues.

The group's combined translation factor is 0.32, which is typical for players coming from the OHL as a whole, but it will prove to be too low for those who simply had later birthdays, like John Tavares (September 20), Taylor Hall (November 14), Patrick Kane (November 19), Gabriel Landeskog (November 23), Matthew Tkachuk (December 11), Connor McDavid (January 13), and Matt Duchene (January 16). In fact, Mitchell Marner and Michael Del Zotto are the only players in the top 12 who spent an extra season in the OHL after being drafted.

The players with late birthdays have an average translation factor of 0.36, which is lower than the 0.41 for the first group but higher than the 0.28 for those of the same age who spent an extra season in the OHL. This further highlights the importance of considering players with late birthdays and that age should be broken all the way down to the month when studying teenage players.

The next group is composed of the players who weren't drafted right away and/or who spent one or two more seasons in the OHL after being drafted. They have an average translation factor of 0.25.

Many of these players eventually developed into solid, high-scoring NHL players, like Mark Scheifele of the Jets, but rarely in their first season. One of the exceptions is Wojtek Wolski, who may have been a bit of a fluke, given that his subsequent scoring totals gradually dropped and that his NHL career was essentially extinguished by age 25. Arizona can only hope that the same fate doesn't befall Max Domi, whose scoring dropped to 38 points in 59 games in 2016–17.

PLAYER	SEASON	GP	G	A	PTS	TEAM	GP	G	A	PTS
Max Domi	2014–15	57	32	70	102	ARI	81	18	34	52
Wojtek Wolski	2005–06	56	47	81	128	COL	76	22	28	50
Robby Fabbri	2014–15	30	25	26	51	STL	72	18	19	37
Mark Scheifele	2012–13	45	39	40	79	WPG	63	13	21	34
Christian Dvorak	2015–16	59	52	69	121	ARI	78	15	18	33
Boone Jenner	2012–13	56	45	37	82	CBJ	72	16	13	29
Sergei Kostitsyn	2006–07	59	40	91	131	MTL	52	9	18	27
Brandon Saad	2011–12	44	34	42	76	CHI	46	10	17	27
Dougie Hamilton	2012–13	32	8	33	41	BOS	64	7	18	25
Wayne Simmonds	2007–08	60	33	42	75	LAK	82	9	14	23
Andrew Shaw	2010–11	66	22	32	54	CHI	37	12	11	23
Chris Tierney	2013–14	67	40	49	89	SJS	43	6	15	21
Brendan Perlini	2015–16	57	25	20	45	ARI	57	14	7	21
Ryan Strome	2012–13	53	34	60	94	NYI	37	7	11	18
Bryan Little	2006–07	57	41	66	107	WPG	48	6	10	16
Kyle Clifford	2009–10	58	28	29	57	LAK	76	7	7	14
Nazem Kadri	2009–10	56	35	58	93	TOR	29	3	9	12
Steve Downie	2006–07	45	35	57	92	PHI	32	6	6	12
Ryan Murphy	2012–13	54	10	38	48	CAR	48	2	10	12
Bobby Ryan	2006–07	63	43	59	102	ANA	23	5	5	10
Darnell Nurse	2014–15	36	10	23	33	EDM	69	3	7	10
Logan Couture	2008–09	62	39	48	87	SJS	25	5	4	9
Nick Foligno	2006–07	66	31	57	88	OTT	45	6	3	9
Cody Ceci	2012–13	69	19	45	64	OTT	49	3	6	9
Vincent Trocheck	2012–13	63	50	59	109	FLA	20	5	3	8
Connor Murphy	2012–13	33	6	12	18	PHX	30	1	7	8
Erik Gudbranson	2010–11	44	12	22	34	FLA	72	2	6	8
Reid Boucher	2012–13	68	62	33	95	NJD	23	2	5	7
John Carlson	2008–09	59	16	60	76	WSH	22	1	5	6
Scott Laughton	2013–14	54	40	47	87	PHI	31	2	4	6
Jamie McGinn	2007–08	51	29	29	58	SJS	35	4	2	6
Nick Paul	2014–15	58	37	29	66	OTT	24	2	3	5
Nick Ritchie	2014–15	48	29	33	62	ANA	33	2	2	4

PLAYER	SEASON	GP	G	A	PTS	TEAM	GP	G	A	PTS
Chris Bigras	2014–15	62	20	51	71	COL	31	1	2	3
Michael McCarron	2014–15	56	28	40	68	MTL	20	1	1	2
Ryan Parent	2006–07	43	3	7	10	PHI	22	0	0	0

Minimum 20 games played in both leagues.

Finally, there are the players who remained in the OHL for the long haul. Their translation factor is only 0.23, since their OHL scoring will be at its peak and truly skilled players are rarely left in the juniors for this long. The one exception is some skilled defencemen, who remain in the juniors for a little longer because they can develop more effectively on a top OHL pairing than they can by fighting for one of the six spots on a strong NHL depth chart, or even in the AHL. Still, it's quite a surprise to see a player break out after a late start the way Alex Pietrangelo did for the St. Louis Blues, injuries or not.

PLAYERS AGED 20 YEARS OR OLDER WHO WENT FROM THE OHL TO THE NHL, 2005–06 TO 2016–17

PLAYER	SEASON	GP	G	A	PTS	TEAM	GP	G	A	PTS
Alex Pietrangelo	2009–10	25	9	20	29	STL	79	11	32	43
Kevin Labanc	2015–16	65	39	88	127	SJS	55	8	12	20
Joseph Blandisi	2014–15	68	52	60	112	NJD	41	5	12	17
Barclay Goodrow	2013–14	63	33	34	67	SJS	60	4	8	12
Ryan Ellis	2010–11	58	24	77	101	NSH	32	3	8	11
Joshua Ho-Sang	2015–16	66	19	63	82	NYI	21	4	6	10
Zack Kassian	2010–11	56	26	51	77	BUF	44	4	6	10
Jared Boll	2006–07	66	28	27	55	CBJ	75	5	5	10
Marc Staal	2006–07	53	5	29	34	NYR	80	2	8	10
Cody Bass	2006–07	53	10	35	45	OTT	21	2	2	4

Minimum 20 games played in both leagues.

And what about the great Connor McDavid? Well, he actually has a January birthday, and thus falls into the age-18 group, with late-birthday players like Kane, Duchene, and Tavares, and he should technically have a 0.36 translation factor. However, that results in an NHLe of 0.92 points per game, or 41 points in 45 games, which is noticeably lower than the 48 he actually achieved. If he had been born a few months sooner, then the 0.41 translation factor used for 17-year-olds would have applied. That translates to 1.05 points per game, which is 47 points in 45 games and within a single point of his actual point total. Once again, even a few months can make all the difference for an accurate translation.

Western Hockey League

The Western Hockey League (WHL) was founded in 1966 as the WCHL, from the Saskatchewan Junior Hockey League (SJHL) and the Canadian Major Junior Hockey League (CMJHL), but it wasn't officially sanctioned by the Canadian Amateur Hockey Association until 1970.

In terms of sheer numbers, the WHL has been providing almost as many players as the OHL recently, with a historically consistent translation factor of 0.29.

Known as a highly physical league throughout the 1970s and 1980s, the WHL still produces some of the harder-hitting and grittier draft picks, as opposed to the higher-scoring talent more commonly found in the two other CHL leagues. That could explain why age doesn't appear to have quite the same impact on a WHL player's scoring as it does on an OHL player. It can also be explained by the lack of data, since only the following two players made the early jump.

PLAYERS AGED 17 WHO WENT FROM THE WHL TO THE NHL, 2005–06 TO 2016–17

PLAYER	SEASON	GP	G	A	PTS	TEAM	GP	G	A	PTS
Ryan Nugent-Hopkins	2010–11	69	31	75	106	EDM	62	18	34	52
Evander Kane	2008–09	61	48	48	96	ATL	66	14	12	26

Minimum 20 games played in both leagues.

Any translation factor specific to the very youngest WHL players will be dominated by Ryan Nugent-Hopkins and his hot rookie NHL season, when he scored 52 points in 62 games, which was double his translation. Given that his NHL points-per-game dropped from that season's 0.84 to 0.6, 0.7, 0.74, 0.62, and 0.52 over the following five seasons, he may simply have been on the right side of some bounces in year one, and his initial 0.42 points-per-game translation may have actually been quite reasonable.

As a further example of the volatile nature of rookie scoring, consider Leon Draisaitl in the upcoming table, who had almost identical WHL scoring stats as Nugent-Hopkins when he was only half a year older, went directly to the same NHL team after being drafted, and yet scored only 0.24 points per game in his rookie season and then jumped to 0.71 and 0.94 points per game in the following two seasons. In essence, some players get a few lucky breaks and others don't, which is exactly why translation factors are calculated in groups.

Obviously, groups need to be a certain size before the lucky and the unlucky start cancelling each other out, but there really haven't been a lot of WHL players to have a big NHL scoring impact at age 19.

Here, Peter Mueller and Tyler Myers are the players who may have enjoyed a few good breaks, but, in my experience, these are more excellent examples of how exceeding expectations by a wide margin in a player's rookie season can signal a decline over the long term. Mueller played only four more NHL seasons, in which he scored 106 points in 208 games before playing his final game at age 24. Myers won the Calder with a 48-point season at age 19 (and a late birthday), but his scoring dropped from 0.59 points per game in 2009–10 to 0.46 and 0.42 over the next two seasons and has been at 0.34 ever since.

PLAYERS AGED 18 WHO WENT FROM THE WHL TO THE NHL, 2005–06 TO 2016–17

PLAYER	SEASON	GP	G	A	PTS	TEAM	GP	G	A	PTS
Peter Mueller	2006–07	51	21	57	78	PHX	81	22	32	54
Tyler Myers	2008–09	58	9	33	42	BUF	82	11	37	48
Morgan Rielly	2012–13	60	12	42	54	TOR	73	2	25	27
Milan Lucic	2006–07	70	30	38	68	BOS	77	8	19	27

PLAYER	SEASON	GP	G	A	PTS	TEAM	GP	G	A	PTS
Seth Jones	2012–13	61	14	42	56	NSH	77	6	19	25
Ryan Johansen	2010–11	63	40	52	92	CBJ	67	9	12	21
Curtis Lazar	2013–14	58	41	35	76	OTT	67	6	9	15
Brett Connolly	2010–11	59	46	27	73	TBL	68	4	11	15
Luke Schenn	2007–08	57	7	21	28	TOR	70	2	12	14
Jake Virtanen	2014–15	50	21	31	52	VAN	55	7	6	13
Leon Draisaitl	2013–14	64	38	67	105	EDM	37	2	7	9
Luca Sbisa	2007–08	62	6	27	33	PHI	39	0	7	7
Colton Gillies	2007–08	58	24	23	47	MIN	45	2	5	7
Brandon Sutter	2007–08	59	26	23	49	CAR	50	1	5	6
James Wright	2008–09	71	21	26	47	TBL	48	2	3	5
Mirco Mueller	2013–14	60	5	22	27	SJS	39	1	3	4
Nino Niederreiter	2010–11	55	41	29	70	NYI	55	1	0	1

Minimum 20 games played in both leagues.

The WHL has produced some legitimately high-scoring NHL players, but most of them required a few more seasons to develop, either in the NHL, like Milan Lucic and Ryan Johansen, or in the WHL, like Jordan Eberle and Jamie Benn.

Technically, Benn was actually drafted out of the BCHL in the fifth round of the 2007 draft, but he then spent two full seasons with the WHL's Kelowna Rockets before breaking out with 22 goals and 41 points as a 20-year-old in his rookie season. His scoring continued to rise, earning him the Art Ross Trophy at age 25. That could foreshadow great success for those with similar stats, like Tampa Bay's Brayden Point.

PLAYERS AGED 19 WHO WENT FROM THE
WHL TO THE NHL, 2005–06 TO 2016–17

PLAYER	SEASON	GP	G	A	PTS	TEAM	GP	G	A	PTS
Leon Draisaitl	2014–15	32	19	34	53	EDM	72	19	32	51
Jordan Eberle	2009–10	57	50	56	106	EDM	69	18	25	43
Sam Reinhart	2014–15	47	19	46	65	BUF	79	23	19	42
Jamie Benn	2008–09	56	46	36	82	DAL	82	22	19	41

PLAYER	SEASON	GP	G	A	PTS	TEAM	GP	G	A	PTS
Brayden Point	2015–16	48	35	53	88	TBL	68	18	22	40
Martin Hanzal	2006–07	60	26	59	85	PHX	72	8	27	35
Ivan Provorov	2015–16	62	21	52	73	PHI	82	6	24	30
Brendan Gallagher	2011–12	54	41	36	77	MTL	44	15	13	28
Travis Hamonic	2009–10	41	11	33	44	NYI	62	5	21	26
Brandon McMillan	2009–10	55	25	42	67	ANA	60	11	10	21
Ryan Murray	2012–13	23	2	15	17	CBJ	66	4	17	21
Gilbert Brulé	2005–06	27	23	15	38	CBJ	78	9	10	19
Brayden Schenn	2010–11	29	22	35	57	PHI	54	12	6	18
Damon Severson	2013–14	64	15	46	61	NJD	51	5	12	17
Matt Dumba	2013–14	26	8	16	24	MIN	58	8	8	16
Brandon Carlo	2015–16	52	4	23	27	BOS	82	6	10	16
Oscar Moller	2007–08	63	39	43	82	LAK	40	7	8	15
Roman Horak	2010–11	64	26	52	78	CGY	61	3	8	11
Sven Baertschi	2011–12	47	33	61	94	CGY	20	3	7	10
Mikael Backlund	2008–09	28	12	18	30	CGY	23	1	9	10
Emerson Etem	2011–12	65	61	46	107	ANA	38	3	7	10
Kris Russell	2006–07	59	32	37	69	CBJ	67	2	8	10
Zach Boychuk	2008–09	43	28	29	57	CAR	31	3	6	9
Cody Eakin	2010–11	56	36	47	83	DAL	30	4	4	8
Cam Barker	2005–06	26	5	13	18	CHI	35	1	7	8
Nic Petan	2014–15	54	15	74	89	WPG	26	2	4	6
Karl Alzner	2007–08	60	7	29	36	WSH	30	1	4	5
Blair Jones	2005–06	72	35	50	85	TBL	20	1	2	3
Alexander Edler	2005–06	62	13	40	53	VAN	22	1	2	3
Jeff Schultz	2005–06	68	7	33	40	WSH	38	0	3	3
Tomas Vincour	2009–10	57	29	19	48	DAL	24	1	1	2

Minimum 20 games played in both leagues.

There are several players with highly impressive WHL totals who never went on to achieve considerable success in the NHL. Many of these players were considered too small, but what about Blair Jones, who is 6-foot-3, 215 pounds, and a reasonably physical player?

Jones may have had a fantastic scoring rate in 2005–06, but it was his fourth season in the WHL. He actually scored 7 points in 37 games in year one, 31 points in 72 games in year two, and 50 points in 68 games in year three, when he was finally drafted. He was a 20-year-old playing against teenagers when his scoring jumped from 26 goals through his first 175 games to 35 in his final 72, so it's really no surprise that his success didn't translate to the NHL.

While this process can help translate a single season, Jones serves as a reminder that an actual projection should be based on several seasons. Naturally, there can still be surprises, like Emerson Etem, whose scoring totals roughly matched Eberle's throughout his three-season WHL career, but who still bounced around three NHL teams, scoring just 46 points in 173 games. Studying multiple seasons may teach us how to keep the scoring totals of prospects like Jones in perspective, but statistically there is still plenty of work to be done to make distinctions between players like Eberle and Etem.

Finally, there's the group of players who are left in the WHL for the long haul. Once again, these are generally defencemen who weren't likely to make a big NHL splash, scoring-wise. Matt Calvert and Devin Setoguchi are the lone forwards. The former has never exceeded 24 points in an NHL season, while the latter had one miracle 31-goal season with Joe Thornton before his career as a top-six forward effectively wrapped up at age 26.

PLAYERS AGED 20 OR OLDER WHO WENT FROM THE WHL TO THE NHL, 2005–06 TO 2016–17

PLAYER	SEASON	GP	G	A	PTS	TEAM	GP	G	A	PTS
Matt Calvert	2009–10	68	47	52	99	CBJ	42	11	9	20
Devin Setoguchi	2006–07	55	36	29	65	SJS	44	11	6	17
Jared Cowen	2010–11	58	18	30	48	OTT	82	5	12	17
Stefan Elliott	2010–11	71	31	50	81	COL	39	4	9	13
Jared Spurgeon	2009–10	54	8	43	51	MIN	53	4	8	12
Luca Sbisa	2009–10	29	4	14	18	ANA	68	2	9	11
Brayden McNabb	2010–11	59	21	51	72	BUF	25	1	7	8
Derrick Pouliot	2013–14	58	17	53	70	PIT	34	2	5	7

Minimum 20 games played in both leagues.

Quebec Major Junior Hockey League

The Quebec Major Junior Hockey League (QMJHL) is the fastest and highest-scoring league in the Canadian Juniors. Founded in 1969 with the merger of the Quebec Junior Hockey League (QJHL) and the Metropolitan Montreal Junior Hockey League (MMJHL), the QMJHL directly provides an average of three NHL regulars per season, with a consistent translation factor of around 0.25. That low factor doesn't indicate that the QMJHL is the lowest-quality league in the CHL, but just that goals are easier to come by.

Just like in the WHL, there really isn't enough data to break down every player by age. For example, there have been only two players who were drafted early and moved directly to the NHL. One is a blue chipper, while the other only played about half of the lockout-shortened 2012–13 season.

PLAYERS AGED 17 WHO WENT FROM THE QMJHL TO THE NHL, 2005–06 TO 2016–17

PLAYER	SEASON	GP	G	A	PTS	TEAM	GP	G	A	PTS
Nathan MacKinnon	2012–13	44	32	43	75	COL	82	24	39	63
Mikhail Grigorenko	2011–12	59	40	45	85	BUF	25	1	4	5

Minimum 20 games played in both leagues.

Unlike the rest of the CHL, the highest-scoring players on the QMJHL's 18-year-old players list don't have late birthdays; they are rookies who were given one extra season in the QMJHL after being drafted. David Perron and Dmitry Kulikov were the only players to go directly to the NHL after being drafted, and they're both in the bottom half of the list.

PLAYERS AGED 18 WHO WENT FROM THE QMJHL TO THE NHL, 2005–06 TO 2016–17

PLAYER	SEASON	GP	G	A	PTS	TEAM	GP	G	A	PTS
Nikolaj Ehlers	2014–15	51	37	64	101	WPG	72	15	23	38
Jakub Voracek	2007–08	53	33	68	101	COL	80	9	29	38

PLAYER	SEASON	GP	G	A	PTS	TEAM	GP	G	A	PTS
Jonathan Drouin	2013–14	46	29	79	108	TBL	70	4	28	32
Jonathan Huberdeau	2011–12	37	30	42	72	FLA	48	14	17	31
Guillaume Latendresse	2005–06	51	43	40	83	MTL	80	16	13	29
David Perron	2006–07	70	39	44	83	STL	62	13	14	27
Sean Couturier	2010–11	58	36	60	96	PHI	77	13	14	27
Marc-Édouard Vlasic	2005–06	66	16	57	73	SJS	81	3	23	26
Anthony Beauvillier	2015–16	47	40	39	79	NYI	66	9	15	24
James Sheppard	2006–07	56	33	63	96	MIN	78	4	15	19
Dmitry Kulikov	2008–09	57	12	50	62	FLA	68	3	13	16

Minimum 20 games played in both leagues.

The other interesting result is how two of the top three players, Jakub Voracek and Jonathan Drouin, actually retained a far higher proportion of their assists (39%) than their goals (21%). Is this just a coincidence? The concept of calculating separate translation factors for goals and assists in the AHL was explored in *Hockey Abstract*, but there simply isn't enough data to do so for the QMJHL, especially since that pattern doesn't apply to Nikolaj Ehlers or the rest of the group.[31]

In 2016–17, Drouin broke out with 53 points in 73 games and Ehlers with 64 points in 82 games. Given their statistical similarity with Voracek, this isn't a surprise. Voracek broke out in his fifth NHL season with 46 points in 48 games at age 23, and he was competing for the league scoring title two years later. Drouin and Ehlers could be among the league scoring leaders in short order as well.

Other than Nathan MacKinnon, the QMJHL player with the greatest immediate NHL success since the 2005 lockout is Michael Frolik. He was actually drafted from Kladno in the Czech league, but he then played two seasons with Rimouski in the QMJHL, where he scored a respectable 138 points in 97 games. In his rookie NHL season (2008–09) at age 20, Frolik scored 21 goals and 45 points with the Florida Panthers. Although he has scored between 42 and 44 points four times since then, that rookie result is his career high, and

---

31. Rob Vollman, "Translating Data from Other Leagues," *Hockey Abstract* (2013), 159–182.

it is therefore not the typical level of scoring to expect from other players with comparable QMJHL scoring totals.

**PLAYERS AGED 19 WHO WENT FROM THE
QMJHL TO THE NHL, 2005-06 TO 2016-17**

PLAYER	SEASON	GP	G	A	PTS	TEAM	GP	G	A	PTS
Michael Frolik	2007–08	45	24	41	65	FLA	79	21	24	45
Alexander Radulov	2005–06	62	61	91	152	NSH	64	18	19	37
Jonathan Huberdeau	2012–13	30	16	29	45	FLA	69	9	19	28
Nikita Kucherov	2012–13	33	29	34	63	TBL	52	9	9	18
Kris Letang	2006–07	40	14	38	52	PIT	63	6	11	17
Charlie Coyle	2011–12	23	15	23	38	MIN	37	8	6	14
Jordan Caron	2009–10	43	26	27	53	BOS	23	3	4	7
Mikhail Grigorenko	2013–14	23	15	24	39	BUF	25	3	3	6
Timo Meier	2015–16	52	34	53	87	SJS	34	3	3	6
Marco Scandella	2009–10	31	9	22	31	MIN	20	0	2	2
Patrice Cormier	2009–10	31	11	20	31	ATL	21	1	1	2
Luc Bourdon	2006–07	36	4	16	20	VAN	27	2	0	2

Minimum 20 games played in both leagues.

If there ever was a post-Crosby QMJHL player for whom there was a great scoring projection, then it is Alexander Radulov. In his second QMJHL season after being drafted out of Russia, Radulov set a number of team scoring records, and his 50-game scoring streak is second in league history to Mario Lemieux, who I hear was pretty good at hockey.

However, Radulov never really found success in the NHL. He followed up his relatively modest rookie season with a solid 26 goals and 58 points in 2007–08, but he then played in (and dominated) Russia's KHL throughout the rest of his 20s. (When he returned to the NHL at age 30, it was interesting to see how many points he scored, but that's a topic for a later section.)

PLAYERS AGED 20 OR OLDER WHO WENT FROM THE QMJHL TO THE NHL, 2005-06 TO 2016-17

PLAYER	SEASON	GP	G	A	PTS	TEAM	GP	G	A	PTS
Claude Giroux	2007–08	55	38	68	106	PHI	42	9	18	27
Louis Leblanc	2010–11	51	26	32	58	MTL	42	5	5	10

Minimum 20 games played in both leagues.

Good grief, what was Claude Giroux still doing in the QMJHL at age 20? And why weren't his scoring totals more dominant? Based on his 103 points in 69 games at age 18 and 112 points in 63 games at age 19, one would have expected at least 135 points in 55 games at age 20, according to Iain's age adjustments. It was obviously an aberration, since his 27 points in 42 NHL games works out to 52 points over an 82-game schedule, which would have made him the second-highest-scoring QMJHL rookie since the lockout. Yet another reason to look at several seasons when attempting to project a player's future.

Western Collegiate Hockey Association

These days, a growing number of NHL rookies come directly from college hockey. The National Collegiate Athletic Association (NCAA) Division I hockey is composed of six different leagues of varying quality, with an average translation factor that's just a little bit lower than Desjardins's original calculation of 0.33.

Historically, the strongest league was certainly the Western Collegiate Hockey Association (WCHA), which was founded in 1959–60 as a continuation of the Western Intercollegiate Hockey League (WIHL). At first, only about a single player per season moved directly to the NHL, usually for only a short trial. Things picked up throughout the 1970s, and by the 1980s there were about half a dozen regular NHLers who had played in the WCHA the previous season, which remained the case over the following decades.

In total, there have been 36 players who have played at least 20 games in both the WCHA and then the NHL since 2005–06, and they've had an unexpectedly high average translation factor of 0.42. Historically, the all-time average translation factor was generally 0.34 right up until the 2004–05 lockout, but it shot up after that.

To muddy the waters, a major realignment took place prior to the 2013–14 season, with six of the best WCHA teams moving to the National Collegiate Hockey Conference (NCHC) and two others moving to the new Big Ten Conference.

While the WCHA technically does still exist, Colton Parayko is the only player to head directly to the NHL from its new iteration and to play at least 20 games. In essence, the WCHA no longer has a translation factor, and its historical data is only useful in understanding the brand-new NCHC and Big Ten conferences.

National Collegiate Hockey Conference

Founded from six members of the WCHA and two teams from the now-defunct Central Collegiate Hockey Association (CCHA), the NCHC began play in the 2013–14 season as arguably the strongest Division I conference.

Given that the old WCHA had a translation factor of 0.42 and the CCHA had 0.33, the NCHC's league translation factor was originally estimated to be around 0.40. Initially, I thought that was optimistic, given that the WCHA's translation factor plunged in 2012–13, right before the realignment, as Justin Schultz, Drew Shore, Jaden Schwartz, and Jason Zucker retained only 0.28 of their scoring. However, based on the six players who have gone directly from the new NCHC to the NHL and played at least 20 games, the translation factor might be higher than the expected 0.4. Thanks in large part to Jake Guentzel, who scored 33 points in 40 games, the average translation factor for these six players is up over 0.45.

Regardless of where the exact translation factor settles for the NCHC, the old WCHA was a very strong conference, as demonstrated by the wealth of extraordinary players in the following table. The new NCHC should be in that same ballpark.

PLAYERS WHO WENT FROM THE OLD WCHA
OR NCHC TO THE NHL, 2005-06 TO 2016-17

AGE	PLAYER	SEASON	GP	G	A	PTS	TEAM	GP	G	A	PTS
20	Paul Stastny	2005–06	39	19	34	53	COL	82	28	50	78
18	Jonathan Toews	2006–07	34	18	28	46	CHI	64	24	30	54
24	Matt Read	2010–11	37	22	13	35	PHI	79	24	23	47
21	Blake Wheeler	2007–08	44	15	20	35	BOS	81	21	24	45
19	Derek Stepan	2009–10	41	12	42	54	NYR	82	21	24	45
21	Matt Carle	2005–06	39	11	42	53	SJS	77	11	31	42
20	Travis Zajac	2005–06	46	18	29	47	NJD	80	17	25	42
21	T.J. Oshie	2007–08	42	18	27	45	STL	57	14	25	39
20	Nick Bjugstad	2012–13	40	21	15	36	FLA	76	16	22	38
21	Craig Smith	2010–11	41	19	24	43	NSH	72	14	22	36
18	Erik Johnson	2006–07	41	4	20	24	STL	69	5	28	33
21	Jake Guentzel	2015–16	35	19	27	46	PIT	40	16	17	33
20	Jake Gardiner	2010–11	41	10	31	41	TOR	75	7	23	30
18	Phil Kessel	2005–06	39	18	33	51	BOS	70	11	18	29
21	Joe Pavelski	2005–06	43	23	33	56	SJS	46	14	14	28
19	Nick Schmaltz	2015–16	37	11	35	46	CHI	61	6	22	28
20	Drew Stafford	2005–06	42	24	24	48	BUF	41	13	14	27
21	Justin Schultz	2011–12	37	16	28	44	EDM	48	8	19	27
20	Matt Niskanen	2006–07	39	9	22	31	DAL	78	7	19	26
21	Troy Stecher	2015–16	43	8	21	29	VAN	71	3	21	24
21	David Backes	2005–06	38	13	29	42	STL	49	10	13	23
19	Justin Faulk	2010–11	39	8	25	33	CAR	66	8	14	22
21	Mason Raymond	2006–07	39	14	32	46	VAN	49	9	12	21
18	Kyle Turris	2007–08	36	11	24	35	PHX	63	8	12	20
20	Jaccob Slavin	2014–15	34	5	12	17	CAR	63	2	18	20
21	Drake Caggiula	2015–16	39	25	26	51	EDM	60	7	11	18
21	Erik Haula	2012–13	37	16	35	51	MIN	46	6	9	15
23	Matt Frattin	2010–11	44	36	24	60	TOR	56	8	7	15
21	Drew Shore	2011–12	42	22	31	53	FLA	43	3	10	13
19	Jaden Schwartz	2011–12	30	15	26	41	STL	45	7	6	13
21	Ryan Potulny	2005–06	41	38	25	63	PHI	35	7	5	12
20	Ryan McDonagh	2009–10	43	4	14	18	NYR	40	1	8	9
21	Blake Geoffrion	2009–10	40	28	22	50	NSH	20	6	2	8

AGE	PLAYER	SEASON	GP	G	A	PTS	TEAM	GP	G	A	PTS
23	Paul LaDue	2015–16	41	5	14	19	LAK	22	0	8	8
22	Steve Wagner	2006–07	38	6	23	29	STL	24	2	6	8
21	Andrej Sustr	2012–13	39	9	16	25	TBL	43	1	7	8
18	Nick Leddy	2009–10	30	3	8	11	CHI	46	4	3	7
21	Nate Schmidt	2012–13	40	9	23	32	WSH	29	2	4	6
22	Jack Hillen	2007–08	41	6	31	37	NYI	40	1	5	6
21	Chris Butler	2007–08	41	3	14	17	BUF	47	2	4	6
20	Jason Zucker	2011–12	38	22	24	46	MIN	20	4	1	5
20	Andreas Nodl	2007–08	40	18	26	44	PHI	38	1	3	4

Minimum 20 games played in both leagues.

Of all the names on this list, the most interesting story is Jonathan Toews, who somehow managed to score 46 points in 34 games at age 18. To dominate Division I hockey as a teenager is simply incredible, and it foreshadowed the instant NHL success he became.

Scrolling further down the list, there are several players who scored at Toews's rate but at an older age and didn't enjoy immediate (or future) NHL scoring success, like Blake Geoffrion, Ryan Potulny, Drew Shore, Matt Frattin, and several others who missed the 20-game cut-off. That's yet more evidence that the NHL translation process is quite different for teenagers than it is for everybody else.

Big Ten

Similarly to the NCHC, the Big Ten conference began play in the 2013–14 season with six teams from the disbanded CCHA, two teams from the WCHA, and one independent team, Penn State. Based on that membership and the translation factors of their predecessor conferences, Big Ten's translation factor can be estimated at 0.35. Nobody moved directly to the NHL and played at least 20 games after its inaugural 2013–14 season, but a notable rookie did so in each of the following seasons, Dylan Larkin and Zach Werenski, both of whom were 19.

PLAYERS WHO WENT FROM THE BIG 10
TO THE NHL, 2013–14 TO 2016–17

AGE	PLAYER	SEASON	GP	G	A	PTS	TEAM	GP	G	A	PTS
18	Zach Werenski	2015–16	36	11	25	36	CBJ	78	11	36	47
18	Dylan Larkin	2014–15	35	15	32	47	DET	80	23	22	45
21	Mike Reilly	2014–15	39	6	36	42	MIN	29	1	6	7
20	Tyler Motte	2015–16	38	32	24	56	CHI	33	4	3	7
19	Kyle Connor	2015–16	38	35	36	71	WPG	20	2	3	5
20	J.T. Compher	2015–16	38	16	47	63	COL	21	3	2	5

Minimum 20 games played in both leagues.

With Larkin and Werenski tugging it one direction and four others tugging it in another direction, the actual translation factor has been 0.32 so far.

Central Collegiate Hockey Association

The CCHA was established in 1971 and disbanded in 2013, but it's important to study it, as it's useful when estimating the translation factor for the new Big Ten and NCHC conferences.

The first immediate jumps to the NHL began in the late 1970s, with about one per season, starting with John Markell, Tom Laidlaw, and Steve Bozek. The numbers gradually picked up and, by the 1990s, there were two or three NHL regulars per season who had spent the previous season in the CCHA, which remained the case until the 2013 realignment.

Throughout history, the average translation factor for the CCHA was consistently just above 0.30, which is just a little bit lower than the overall collegiate average.

While these players don't have the same pedigree as those who came from the WCHA, there is a surprising number of solid top-four defencemen, and there appears to be a reasonable quantity of effective secondary talent up front.

PLAYERS WHO WENT FROM THE CCHA
TO THE NHL, 2005-06 TO 2013-14

AGE	PLAYER	SEASON	GP	G	A	PTS	TEAM	GP	G	A	PTS
19	Andrew Cogliano	2006–07	38	24	26	50	EDM	82	18	27	45
22	Carl Hagelin	2010–11	44	18	31	49	NYR	64	14	24	38
18	Jacob Trouba	2012–13	37	12	17	29	WPG	65	10	19	29
22	Danny Dekeyser	2012–13	35	2	13	15	DET	65	4	19	23
21	Dan Sexton	2008–09	38	17	22	39	ANA	41	9	10	19
23	Ryan Jones	2007–08	42	31	18	49	NSH	46	7	10	17
22	Anders Lee	2012–13	41	20	18	38	NYI	22	9	5	14
22	Bill Thomas	2005–06	41	27	23	50	PHX	24	8	6	14
21	T.J. Hensick	2006–07	41	23	46	69	COL	31	6	5	11
19	Max Pacioretty	2007–08	37	15	24	39	MTL	34	3	8	11
20	Jon Merrill	2012–13	21	2	9	11	NJD	52	2	9	11
20	Jack Johnson	2006–07	36	16	23	39	LAK	74	3	8	11
21	Kevin Porter	2007–08	43	33	30	63	PHX	34	5	5	10
21	Steven Kampfer	2009–10	45	3	23	26	BOS	38	5	5	10
21	David Booth	2005–06	37	13	22	35	FLA	48	3	7	10
20	Reilly Smith	2011–12	39	30	18	48	DAL	37	3	6	9
22	Christian Hanson	2008–09	37	16	15	31	TOR	31	2	5	7
23	Andy Greene	2005–06	39	9	22	31	NJD	23	1	5	6
22	Jeff Petry	2009–10	38	4	25	29	EDM	35	1	4	5
20	Ian Cole	2009–10	30	3	16	19	STL	26	1	3	4

Minimum 20 games played in both leagues.

The results can be quite unpredictable. Players like Kevin Porter, Bill Thomas, Ryan Jones, Dan Sexton, and T.J. Hensick fought for scoring titles in the CCHA, but they didn't achieve the same success in the NHL. As always, the key is to base a translation on an entire group and then to drill down on individual players using other means.

ECAC Hockey was founded in 2004 and was previously affiliated with the Eastern College Athletic Conference, which was formed in 1961 in the northeastern United States. The 2013 realignment did not affect ECAC, so its historical 0.23 translation factor probably remains valid.

Why "probably"? Since the 2005–06 season, ECAC has only produced nine players who immediately went on to the NHL, and Jeff Halpern was the only immediate NHL regular in the 10 years preceding that. Until Jimmy Vesey, there hadn't been an immediate NHL regular since Alexander Killorn in 2011–12.

PLAYERS WHO WENT FROM THE ECAC
TO THE NHL, 2005-06 TO 2016-17

AGE	PLAYER	SEASON	GP	G	A	PTS	TEAM	GP	G	A	PTS
22	Jimmy Vesey	2015–16	33	24	22	46	NYR	80	16	11	27
22	Alexander Killorn	2011–12	34	23	23	46	TBL	38	7	12	19
24	Jesse Winchester	2007–08	40	8	29	37	OTT	76	3	15	18
23	Colin Greening	2009–10	34	15	20	35	OTT	24	6	7	13
23	Nick Lappin	2015–16	31	17	16	33	NJD	43	4	3	7
22	David Jones	2006–07	33	18	26	44	COL	27	2	4	6
23	Harry Zolnierczyk	2010–11	30	16	15	31	PHI	37	3	3	6
23	John Zeiler	2005–06	28	13	15	28	LAK	23	1	2	3
23	Tanner Glass	2006–07	32	8	20	28	FLA	41	1	1	2

Minimum 20 games played in both leagues.

There's no way to study the impact of a player's age in the ECAC, since the youngest player to move directly to the NHL (and play 20 games) was already 22 years old.

In terms of quality, ECAC ranks below the NCHC, Big Ten, and Hockey East conferences, but it is arguably ahead of the new WCHA and the Atlantic conference, the latter of which hasn't produced any immediate NHL players.

In 1984, ECAC's New England–based Ivy League teams split off to form the Hockey East Association. Since then, it has been a reliable source of secondary NHL talent, as well as some top players like Johnny Gaudreau, Jack Eichel, Kevin Hayes, Kevin Shattenkirk, and James van Riemsdyk.

Since 2005–06, the following 31 players have moved from Hockey East to the NHL and played at least 20 games. Their average translation factor is 0.38, which has been boosted by two of the more recent high-scoring names on the list, Gaudreau and Eichel.

PLAYERS WHO WENT FROM HOCKEY EAST TO THE NHL, 2005-06 TO 2016-17

AGE	PLAYER	SEASON	GP	G	A	PTS	TEAM	GP	G	A	PTS
20	Johnny Gaudreau	2013–14	40	36	44	80	CGY	80	24	40	64
18	Jack Eichel	2014–15	40	26	45	71	BUF	81	24	32	56
21	Kevin Hayes	2013–14	40	27	38	65	NYR	79	17	28	45
21	Kevin Shattenkirk	2009–10	38	7	22	29	COL	72	9	34	43
19	James van Riemsdyk	2008–09	36	17	23	40	PHI	78	15	20	35
21	Ben Hutton	2014–15	39	9	12	21	VAN	75	1	24	25
19	Matt Nieto	2012–13	39	18	19	37	SJS	66	10	14	24
17	Noah Hanifin	2014–15	37	5	8	13	CAR	79	4	18	22
22	Bobby Butler	2009–10	39	29	24	53	OTT	36	10	11	21
22	Torrey Mitchell	2006–07	39	12	23	35	SJS	82	10	10	20
23	Brandon Yip	2008–09	45	20	23	43	COL	32	11	8	19
20	Brett Pesce	2014–15	31	3	13	16	CAR	69	4	12	16
19	Colin Wilson	2008–09	43	17	38	55	NSH	35	8	7	15
21	Matt Benning	2015–16	41	6	13	19	EDM	62	3	12	15
24	Matt Gilroy	2008–09	45	8	29	37	NYR	69	4	11	15
21	Cam Atkinson	2010–11	39	31	21	52	CBJ	27	7	7	14
23	Viktor Stalberg	2008–09	39	24	22	46	TOR	40	9	5	14
22	Mark Fayne	2009–10	34	5	17	22	NJD	57	4	10	14
23	Brian Flynn	2011–12	40	18	30	48	BUF	26	6	5	11
22	Justin Braun	2009–10	36	8	23	31	SJS	28	2	9	11
20	Frank Vatrano	2014–15	36	18	10	28	BOS	39	8	3	11

AGE	PLAYER	SEASON	GP	G	A	PTS	TEAM	GP	G	A	PTS
22	Christian Folin	2013–14	41	6	14	20	MIN	40	2	8	10
21	Jimmy Hayes	2010–11	39	21	12	33	CHI	31	5	4	9
21	Zach Sanford	2015–16	41	13	26	39	STL	39	4	4	8
22	Mike Lundin	2006–07	40	6	14	20	TBL	81	0	6	6
21	Stéphane Da Costa	2010–11	33	14	31	45	OTT	22	3	2	5
21	Benn Ferriero	2008–09	37	8	18	26	SJS	24	2	3	5
20	Chris Kreider	2011–12	44	23	22	45	NYR	23	2	1	3
21	Josh Manson	2013–14	33	3	7	10	ANA	28	0	3	3
22	Chad Ruhwedel	2012–13	41	7	16	23	BUF	21	0	1	1
21	Nick Bonino	2009–10	33	11	27	38	ANA	26	0	0	0

Minimum 20 games played in both leagues.

While the CHL and NCAA Division I provide many of the NHL's best rookies, the majority of players fight their way in by spending at least one season in one of the veteran leagues, like the AHL, or one of the stronger European leagues. Let's dig into those leagues.

Veterans

The process for translating scoring data from the world's best professional leagues to the NHL is the same for veterans as it is for prospects, but without the player's age having as significant an impact. The end result is that we can more easily chart a player's career arc and integrate it into a projection, even if the player spent one or more years in Europe or the AHL.

Translated data will never be perfect. Even within the NHL, a player's scoring rate will bounce up and down because of changes to his luck, coaching, ice time, linemates, opponents, and sometimes even the team for which he plays. For a player entering the NHL from another league, almost all of these factors will be completely changing—and yet translating his non-NHL data using a simple ten-year-old system can actually help predict his NHL scoring data over half as well as actual NHL data. It gets even better if more recent advances are used, like separating goals and assists, factoring in things like age, ice time, manpower, and scoring level, or making adjustments based on similar players who recently made the same transition.

At the beginning of this chapter, we used Panarin's statistics as an example of how to translate data from the KHL to the NHL, and we discussed how reliable the results can be. Using that same approach, there could be some excitement that Vadim Shipachyov of the Vegas Golden Knights could perform even better. His 76 points in 50 games in the KHL in 2016–17 translates to 80 points in 68 games in the NHL. But is that a realistic expectation?

First of all, Shipachyov is 30 years old and will be playing for an expansion team that is unlikely to include the same calibre of linemate that Panarin enjoyed, Chicago's Patrick Kane. In fact, it probably won't even include someone of the same calibre that he played with in St. Petersburg, Ilya Kovalchuk.

Secondly, his strong 2016–17 season was not typical for Shipachyov. Just like in the NHL, sometimes players can get into the right situation and have a one-time career season. By translating his entire career, and not focusing solely on his final KHL season, it's more obvious that expectations should be in the low 60s by NHL standards (assuming he stays healthy). One of the advantages of taking a year to go to print is that by the time you read this, you'll know how well his 2017–18 scoring results fit in with the following translated career data (on the right).

VADIM SHIPACHYOV'S NHL-EQUIVALENT CAREER SCORING[32]

AGE	TEAM	GP	G	A	PTS	SEASON	GP	G	A	PTS
21	Cherepovets	29	4	5	9	2008–09	42	4	5	9
22	Cherepovets	55	14	30	44	2009–10	81	14	31	45
23	Cherepovets	51	13	25	38	2010–11	77	14	26	40
24	Cherepovets	54	22	37	59	2011–12	82	24	40	64
25	Cherepovets	51	17	23	40	2012–13	47	11	15	26
26	St. Petersburg	52	12	20	32	2013–14	79	14	23	37
27	St. Petersburg	49	12	42	54	2014–15	67	12	41	53

---

32. Raw data for the NHL translations from Hockey DB, accessed July 15, 2017, http://www.hockeydb.com.

AGE	TEAM	GP	G	A	PTS	SEASON	GP	G	A	PTS
28	St. Petersburg	54	17	43	60	2015–16	74	17	44	61
29	St. Petersburg	50	26	50	76	2016–17	68	27	52	79

If he had been playing in the NHL his whole career, then Shipachyov would have scored an estimated 137 goals, 277 assists, and 414 points in 617 games. With stats like that, a player like Mats Zuccarello comes to mind, or perhaps a younger Ales Hemsky. That comparison doesn't preclude an 80-point season, but it makes 60 points a far more realistic expectation.

Seeing all of Shipachyov's career data also serves as an example of how NHL translations allow players to be included in a team's full-fledged projection model, like the one covered in the opening chapter of *Stat Shot*. The translations themselves are not projections, but several seasons' worth of data is a key component in a long-term statistical projection.

Let's start the more detailed examination of each professional league with Shipachyov's former home, the KHL.

Kontinental Hockey League

Though it has been on a steady decline for the past decade, the second-strongest hockey league in the world is still the Kontinental Hockey League (KHL).

The top Russian hockey league has come in many different forms and has had several different names over the years, but it has always been the highest level of hockey outside North America. In fact, the lower-scoring nature of the league, along with a few number-skewing superstars, has resulted in a translation factor that, at its peak just over a decade ago, occasionally flirted with NHL parity.

Through its various incarnations from around 1996 until the 2005 NHL lockout season, including most notably the formation of the Russian Superleague (RSL) in 1999, 48 players moved from the highest Russian hockey league to the NHL, or about six per season, with an average translation factor of 0.88.

During the 2004–05 NHL lockout, a total of 46 players chose to play in the RSL before returning to the NHL, with an average translation factor of 1.04. Because of high-scoring superstars like Jaromir Jagr, Alexander Ovechkin, Ilya Kovalchuk, Pavel Datsyuk, Vincent Lecavalier, and Alexei Kovalev, the average RSLer actually enjoyed a scoring *boost* when returning to the NHL in 2005–06.

At its peak, in the three years between that lockout and the 2008–09 KHL rebranding, 35 players moved from the Russian Superleague to the NHL (about a dozen per season), with a near-parity translation factor of 0.98 among the following players, who played at least 20 games in both leagues. Even if the boosting effect of Evgeni Malkin and Alexander Semin is removed (for whatever arbitrary reason), the translation factor is still a very competitive 0.89.

PLAYERS WHO WENT FROM THE RSL TO THE NHL, 2005–06 TO 2008–09[33]

AGE	PLAYER	SEASON	GP	G	A	PTS	TEAM	GP	G	A	PTS
19	Evgeni Malkin	2005–06	46	21	26	47	PIT	78	33	52	85
21	Alexander Semin	2005–06	42	8	11	19	WSH	77	38	35	73
23	Anton Babchuk	2007–08	57	9	17	26	CAR	72	16	19	35
21	Nikolai Kulemin	2007–08	57	21	12	33	TOR	73	15	16	31
26	Denis Arkhipov	2005–06	50	8	8	16	CHI	79	10	17	27
26	Karel Rachunek	2005–06	45	11	16	27	NYR	66	6	20	26
24	Nikita Alexeev	2005–06	40	6	3	9	TBL/ CHI	78	12	11	23
24	Mark Giordano	2007–08	50	4	8	12	CGY	58	2	17	19
23	Alexander Svitov	2005–06	32	3	6	9	CBJ	76	7	11	18
23	Denis Grebeshkov	2006–07	47	8	9	17	EDM	71	3	15	18
19	Viktor Tikhonov	2007–08	43	7	5	12	PHX	61	8	8	16
24	Evgeny Artyukhin	2007–08	23	3	5	8	TBL	73	6	10	16
22	Stanislav Chistov	2005–06	47	11	21	32	ANA/ BOS	61	5	8	13

33. Raw data from Hockey Reference, accessed July 1, 2013, http://www.hockey-reference.com.

AGE	PLAYER	SEASON	GP	G	A	PTS	TEAM	GP	G	A	PTS
27	Jan Hejda	2005–06	50	3	12	15	EDM	39	1	8	9
25	Alexei Semenov	2006–07	20	1	2	3	SJS	22	1	3	4

Minimum 20 games played in both leagues.

As for the 2012–13 NHL lockout, a total of 37 players competed in the KHL before returning to the NHL that same season and/or the next, including Malkin, Datsyuk, and Ovechkin once again. Among those with at least 20 games in each league, the translation factor was 0.78, which was actually far more typical of the league quality at the time and since.

That was really the first signal that those using Desjardins's original translation factors from 2004 or even 2006 would no longer produce accurate results. As we will discover with other leagues, like Finland's SM-liiga, it is very important to use up-to-date translation factors.

Since its relaunch in 2008–09 as the KHL, 55 players have moved straight to the NHL, or about eight per season, not counting the lockout. Among the following 34 players to have competed in at least 20 games in both leagues, the average translation factor is 0.77, and it's trending down.

PLAYERS WHO WENT FROM THE KHL
TO THE NHL, 2009-10 TO 2016-17

AGE	PLAYER	SEASON	GP	G	A	PTS	TEAM	GP	G	A	PTS
23	Artemi Panarin	2014–15	54	26	36	62	CHI	80	30	47	77
38	Jaromir Jagr	2010–11	49	19	31	50	PHI	73	19	35	54
29	Alexander Radulov	2015–16	53	23	42	65	MTL	76	18	36	54
27	Kyle Wellwood	2010–11	25	5	3	8	WPG	77	18	29	47
26	Jori Lehtera	2013–14	48	12	32	44	STL	75	14	30	44
34	Petr Sykora	2010–11	28	8	8	16	NJD	82	21	23	44
26	Jiri Hudler	2009–10	54	19	35	54	DET	73	10	27	37
21	Evgeny Kuznetsov	2013–14	31	8	13	21	WSH	80	11	26	37
24	Nikita Zaitsev	2015–16	46	8	18	26	TOR	82	4	32	36
25	Anton Babchuk	2009–10	49	9	13	22	CAR/CGY	82	11	24	35

AGE	PLAYER	SEASON	GP	G	A	PTS	TEAM	GP	G	A	PTS
22	Linus Omark	2009–10	56	20	16	36	EDM	51	5	22	27
27	Leo Komarov	2013–14	52	12	22	34	TOR	62	8	18	26
21	Jiri Sekac	2013–14	47	11	17	28	MTL/ANA	69	9	14	23
25	Nikolai Zherdev	2009–10	52	13	26	39	PHI	56	16	6	22
19	Marko Dano	2013–14	41	3	2	5	CBJ	35	8	13	21
23	Alex Burmistrov	2014–15	53	10	16	26	WPG	81	7	14	21
20	Pavel Buchnevich	2015–16	18	4	4	8	NYR	41	8	12	20
20	Vladimir Tarasenko	2011–12	54	23	24	47	STL	38	8	11	19
19	Dmitry Orlov	2010–11	45	2	11	13	WSH	60	3	16	19
26	Roman Cervenka	2011–12	54	23	16	39	CGY	36	9	8	17
23	Sergey Kalinin	2014–15	58	12	13	25	NJD	78	8	7	15
32	Evgeny Medvedev	2014–15	43	3	13	16	PHI	45	4	9	13
27	André Benoit	2011–12	53	5	12	17	OTT	33	3	7	10
23	Nikita Nikitin	2009–10	43	4	9	13	STL	41	1	8	9
25	Leo Komarov	2011–12	46	11	13	24	TOR	42	4	5	9
25	Michal Kempny	2015–16	59	5	16	21	CHI	50	2	6	8
19	Nikita Filatov	2009–10	26	9	13	22	CBJ	23	0	7	7
25	Alexei Emelin	2010–11	52	11	15	26	MTL	67	3	4	7
26	Philip Larsen	2015–16	52	11	25	36	VAN	26	1	5	6
24	Roman Lyubimov	2015–16	52	7	7	14	PHI	47	4	2	6
26	Viktor Tikhonov	2014–15	49	8	16	24	CHI/ARI	50	3	3	6
31	Antti Miettinen	2011–12	20	2	6	8	WPG	22	3	2	5
26	Mark Popovic	2008–09	52	8	15	23	ATL	37	2	2	4
24	Sergei Plotnikov	2014–15	56	15	21	36	PIT/ARI	45	0	3	3

Minimum 20 games played in both leagues, excluding the 2012–13 season.

On an individual basis, it's not always easy to calculate accurate projections. As demonstrated by the spread in the following chart, individual players who score between 0.3 and 0.7 points per game in the KHL (horizontal axis) can produce anywhere from 0.1 to 0.5 points per game in the NHL (vertical axis).

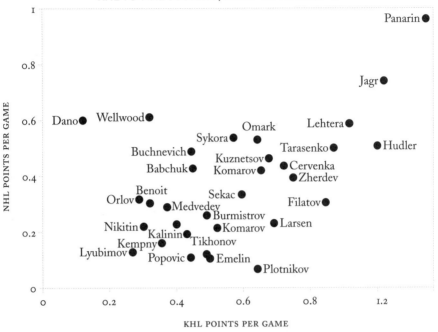

There is obviously a bit of a spread, but an individual player's scoring will ebb and flow regardless of whether he changes leagues or not. Even Shipachyov's scoring in the KHL has fluctuated between 0.62 and 1.52 over the past four seasons, and that's at the relatively stable age range of 26 to 29 and without changing teams.

It's unrealistic to expect these points to all conform to a straight line, but it's possible to see a rough trend line that rises from the bottom left to the top right, running roughly between players like Nikitin and Medvedev and then through Komarov, Cervenka, and Tarasenko. (It helps if you squint.)

It might be easier to see the trend if we consider players together in groups, rather than individuals. If we divide the 34 players into four groups of seven players and one group of six based on their KHL scoring rates, then we get the following bar chart. The scoring progression is now a little more obvious.

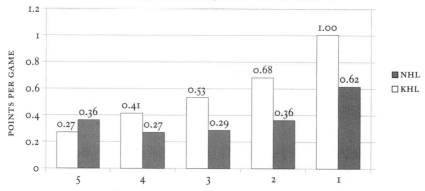

KHL TO NHL SCORING, 2009-10 TO 2016-17

Group 5 might have been skewed by selection bias, as well as by Marko Dano and Kyle Wellwood, and the progression might be more of a gradual curve than a straight line, but we'll revisit some of these quirks in due course. The key points are that the relationship between KHL and NHL scoring is visible, and it's measurable.

Grouping data together is a great way to enlarge sample sizes and to see the overall pattern. Instead of grouping several different players, we can group several seasons of the same player. On the KHL side, we could group each of a player's past three seasons, but since several of these players had only a single KHL season, our already small group of 34 players would get even smaller.

Instead, we can group together a player's data on the NHL side. That would certainly help in the case of Dano and Wellwood. They may have scored 0.6 and 0.61 points per game in their NHL season that immediately followed their tenure in the KHL, but they followed that up with 0.29 (21 points in 72 games) and 0.38 (15 points in 39 games). Those results would place them much closer to the general trend line.

To take more interesting examples, consider Jiri Hudler and Jaromir Jagr. These are two established, veteran players who had virtually identical KHL scoring statistics in consecutive seasons, but Jagr had far more success upon his NHL return.

It may appear that the translation factors failed us, but this discrepancy has worked itself out since then. Jagr has scored 0.73 points per game over the following five NHL seasons combined, which is roughly the same as Hudler's 0.69. In essence, Hudler just had one bad season upon his immediate return and Jagr didn't.

JAGR'S AND HUDLER'S POINTS PER GAME[34]

PLAYER	KHL	TRANSLATED	1ST NHL YEAR	MULTIPLE NHL
Jaromir Jagr	1.02	0.82	0.74	0.73
Jiri Hudler	1.00	0.80	0.51	0.69

If the Hudler/Jagr miscalculation calls anything into question, it's the translation factor itself. Since they were both point-a-game players in the KHL, shouldn't their scoring have levelled off at the KHL translation factor of 0.77 points per game instead of a handful of points below that mark?

One of our many new insights is that a straight-up translation factor can overestimate higher-scoring players. That's also why group 1 in the earlier chart had an average of 1.00 points per game in the KHL but only 0.62 in the NHL.

High-scoring players like Jagr and Hudler aren't always guaranteed the big minutes on the power play and with the top line that they enjoyed in Russia. They may struggle to retain as much of their scoring as middling players, like Leo Komarov, who didn't have much top-line or power-play time in the KHL. As such, their scoring will drop.

On the other hand, lower-scoring players are going to outperform a straight-up translation factor. In any league, the lower-scoring players are often defensive-minded defencemen, who really don't have a lot of scoring to lose when they go to the NHL.

There's also selection bias. It's not like all low-scoring KHL players get NHL opportunities; only the truly exceptional ones are selected. Like the defencemen, they don't have a lot of scoring to lose, nor are they likely to have any top-line or power-play minutes to lose. And if they fail to score at least 0.2 points per game, then they'll likely be cut before they even play 20 games, and therefore won't be selected for our list. So there's a double whammy of selection bias at play.

---

34. Raw data for the NHL translations from Hockey DB, accessed July 15, 2017, http://www.hockeydb.com.

A straight-up translation factor remains fine for the average player, but it will underestimate low-scoring players and overestimate those at the top. The solution may be to assume a base scoring rate higher than 0 points per game, like perhaps 0.25 based on the five-group chart, and then to calculate the translation factor from there, with a far gentler slope. There may not be enough data to work this out with the KHL, but we can explore this concept further using the AHL.

American Hockey League

The beauty of the American Hockey League (AHL) is its abundance of data. For example, 149 players competed in at least 20 AHL games in 2015–16 before playing at least 20 at the NHL level in 2016–17. Multiply that by 10 non-lockout seasons since 2005–06, and you realize there's a huge body of data on which you can conduct more granular analysis, such as separating goals and assists and adjusting for age, both of which were covered in the original *Hockey Abstract*.[35] You can also study different scoring rates and account for ice time and special teams, which we'll explore here.

The major downside of working with AHL data is the limited choice of statistics. The only stats to work with are basic ones like goals, assists, plus/minus, penalty minutes, and shots, plus every player's scoring broken down by manpower situation. There's no ice time, no real-time scoring stats, and certainly no shot-based possession metrics.

As limited as the AHL data set may be, Finland and Sweden are really the only leagues with a greater variety of statistics from which to choose. Unfortunately, there just aren't enough players coming to the NHL from those leagues on which we can base a good study. This is especially true of statistics that weren't made available until recently, like time on ice broken down by manpower situation, which was only introduced to SM-liiga in the 2014–15 season.

---

35. Rob Vollman, "Translating Data from Other Leagues," *Hockey Abstract* (2013), 159–182.

AVAILABLE STATISTICS BY LEAGUE, AS OF 2016–17

LEAGUE	BASIC STATS	SPECIAL TEAMS SCORING	TOTAL ICE TIME	SPECIAL TEAMS ICE TIME	RTSS	CORSI SAT
SHL	Yes	Yes	Yes	Yes	Yes	No
SM-liiga	Yes	Yes	Yes	Yes	Yes	No
AHL	Yes	Yes	No	No	No	No
CHL	Yes	Yes	No	No	No	No
KHL	Yes	No	Yes	No	No	No
NCAA Div. I	Yes	No	No	No	No	No
NLA	Yes	No	No	No	No	No

Fortunately, I'm no stranger to a lack of statistics. I come from a time when we had to create our own. I remember having to cope without NHL save percentage data until 1983–84. I toiled for decades without individual player ice time broken down by manpower situation, and I even remember when *Total Hockey* came out at the end of the 90s, and we finally had team goals scored for and against when players were on the ice.[36] (In short, I have worked with stone-age tools before, and I can do it again. In fact, I relish the opportunity, and I hope that you do too.)

Armed with that experience, there are a few simple things we can do with the basic statistics made available by the AHL. For starters, we can continue our examination of the impact of a player's scoring level on his translation factor.

In general, the AHL translation factor hovers between 0.45 and 0.47. In the original *Hockey Abstract*, we observed that the translation factor was as high as 0.56 for players whose AHL scoring rate was less than half a point a game, but that it dropped to 0.34 for the point-a-gamers.[37] This is especially true when you don't separate players by position, since the latter group includes very few defencemen, whose NHL job usually doesn't depend on their scoring.

Instead of putting hundreds of AHLers on a single chart like

---

36. James Duplacey and Dan Diamond, *Total Hockey: The Official Encyclopedia of the National Hockey League* (New York: Total Sports, 1998).

37. Rob Vollman, "Translating Data from Other Leagues," *Hockey Abstract* (2013), 163.

we did for the KHL, consider the following simplified version. In it, almost 400 AHL players are placed into 16 groups of equal size, based on their per-game scoring rate. As you can see, both with and without the super-imposed trend line, there's a much better option than the standard 0.47 translation factor anchored at zero.

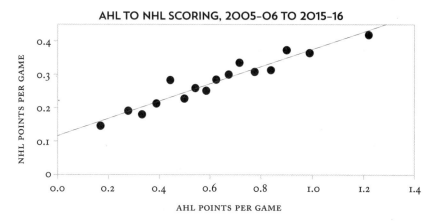

AHL TO NHL SCORING, 2005–06 TO 2015–16

NHL POINTS PER GAME (y-axis)

AHL POINTS PER GAME (x-axis)

According to the trend line, it would be more accurate to assume a baseline scoring rate of 0.11 points per game in the NHL, plus a translation factor of 0.26 multiplied by the player's AHL scoring rate. That will boost the translation for defencemen and lower-scoring forwards appropriately and temper the translations for the high scorers, but it will also leave the vast majority of translations essentially untouched.

We could construct similar charts for all of the leagues discussed in this chapter, although it's hard to have any confidence in the dramatically smaller sets of data. If absolutely forced to repeat this exercise for every league, the following factors might result in more accurate translations for particularly low-scoring or high-scoring players.

NHL TRANSLATION FACTORS, AS OF 2016–17

LEAGUE	BASE	FACTOR
Kontinental Hockey League (KHL)	0.11	0.45
Swedish Hockey League (SHL)[38]	0.22	0.20
American Hockey League (AHL)	0.11	0.26

38. We'll explore shortly why this translation factor really doesn't work for the Swedish Hockey League.

LEAGUE	BASE	FACTOR
NCAA Division I	0.08	0.32
Switzerland (NLA)	0.00	0.40
Finland SM-liiga	0.05	0.30
Canadian Hockey League (CHL)	0.06	0.23

While handy, this approach is a bit of a hack. Generally, the easiest way to get to the root of a problem like this is to ask why it exists. So why does the AHL trend line look like this? When going to the NHL, why do lower-scoring players retain more of their scoring than higher-scoring players?

One reason why it's harder for high-scoring players to retain their scoring when going from other leagues to the NHL is that they lose a lot of their ice time on the power play. A gifted AHL player is likely leading his team with three minutes per game with the man advantage, but he might be a secondary option in the NHL, at best.

Islanders fans might remember Keith Aucoin and Jeff Tambellini, for example. Aucoin scored 70 points in 43 games for the AHL's Hershey Bears in 2011–12 and then just 12 points in 41 games for the Isles the following season. Tambellini managed 76 points in 57 games for the Bridgeport Sound Tigers in 2007–08, but he scored just 15 points in 65 games the following season in the NHL. Since 44 of Aucoin's 70 points came with the man advantage, as did 38 of Tambellini's 76, their modest NHL scoring totals were predictable even without using a modified trend line.

The good news is that the AHL is one of the few leagues that publishes every player's scoring statistics broken down by manpower situation, so we can figure out which players had their scoring boosted by the power play and make the necessary adjustments. The bad news is that studying even-strength scoring by itself doesn't actually lead to more accurate estimates, as shown in the following table.

SEASON-TO-SEASON SCORING RATE
CORRELATION, 2005–06 TO 2014–15

SITUATION	CORRELATION
Comparing a player's overall scoring from one NHL season to the next	0.84
Comparing a player's even-strength scoring from one NHL season to the next	0.76
Comparing a player's overall scoring in the AHL to his overall scoring the following season in the NHL	0.47
Comparing a player's even-strength scoring in the AHL to his even-strength scoring the following season in the NHL	0.47

Minimum 40 games played in each season.

The key lesson here is that setting aside data always comes at a price. The remaining data may be more relevant, but there's less of it, which is why a player's overall scoring has a stronger year-to-year correlation than a player's even-strength scoring, even when dealing exclusively with NHL data.

The other reason the correlation isn't higher is because it's almost impossible to predict which players will have their power-play opportunities cut when they make the NHL and by how much. After all, wouldn't it have been reasonable for the Islanders to try Aucoin and Tambellini on the power play, given their incredible success in those roles in the AHL?

Without the benefit of hindsight, it made sense to expect Aucoin and Tambellini to be assigned more than 0:52 power-play minutes per game in the NHL and, therefore, to have scored more points. However, that doesn't mean we should heave a sigh of despair and resign ourselves to the trend-line hack, since there is some predictive value in knowing how much of a player's scoring occurred at even-strength or with the man advantage.

Let's calculate translation factors separately for even-strength scoring and the power play. Instead of multiplying a player's overall scoring by the standard 0.47 AHL translation factor, we divide overall scoring by manpower situation, multiply by a translation factor of 0.53 for even-strength scoring and 0.21 for the power play, and then add those results back together.

Continuing the preceding example, Aucoin's 0.60 even-strength points per game becomes 0.32 in the NHL, his 1.02 power-play points per game becomes 0.21, which adds up to a final translation of 0.53 points per game. Yes, that also happens to be roughly the same result we get using the trend-line hack, but at least there's now a method to our madness.

Of course, success in the world of hockey analytics means continuing to ask why long after others have stopped. In this case, why is the translation factor so much lower for power-play scoring (0.21) than it is for even-strength scoring (0.53)? Sadly, there's no way to answer that question. Since the AHL doesn't have ice-time data broken down by manpower situation, it's impossible to determine the extent to which the lower translation factor for power-play scoring is caused by decreased opportunities in the NHL or by a lower scoring rate. Indeed, it's even impossible to figure that out at even strength.

Fortunately, we do have the necessary ice-time data for other leagues, like the Swedish Hockey League all the way back to the 2006–07 season. Based on the far smaller sample size of the 36 players who played at least 20 games in the SHL before playing 20 games in the NHL (and not counting the 2012–13 season, due to the skewing effect of the NHL lockout), the average even-strength amount of ice time of each player only dropped from 13.5 minutes per game to 13.2 when going to the NHL, but power-play ice time was cut almost in half, from 2.95 to 1.55. (Although it doesn't really affect scoring, it's still interesting to note that penalty-killing ice time also dropped, from 1.8 minutes per game in the SHL to 1.0 in the NHL.)

As for scoring rates, these 34 players averaged 1.51 points per 60 minutes at even strength in the SHL and then just 1.17 points per 60 minutes in the NHL, for a translation factor of 0.78. On the power play, the average scoring rate dropped from 4.52 points per 60 minutes to 2.96, for a translation factor of 0.66.

Roughly speaking, the two-fifths drop in overall scoring when players move from the SHL to the NHL is due to a one-half drop in power-play opportunities, a one-third drop in individual scoring rates on the power play, and a one-fifth drop in even-strength scoring rate.

That means that players who didn't work the power play in Sweden won't lose nearly as much of their scoring when they come to the NHL as those who did. And even those who continue to work the power play just as regularly in the NHL will still see a drop in their scoring. While we will never know without the necessary data, this rough pattern could apply to the AHL and other leagues as well.

Having already crawled partway into this particular rabbit hole, let's move on and dive into the Swedish league in a little more detail.

Swedish Hockey League

Unlike the KHL, Europe's second-strongest league has become somewhat of a development league for the NHL, albeit one that is still slightly stronger and more competitive than the AHL. Very few established NHLers spent the most recent lockout in Sweden, and very few players older than 23 years old are making the direct jump to the NHL these days.

Among European leagues, Sweden has one of the longest and richest histories with the NHL. Sweden's highest leagues have been a prime source of NHL talent all the way back to Borje Salming over four decades ago. From the time the Swedish Hockey League (formerly Elitserien) was officially formed in the 1975–76 season until the 2005 NHL lockout, a total of 175 players made the move, with an average translation factor of 0.66, based on the 117 players to play in at least 20 games in both leagues.

As for the 2005 lockout year itself, an incredible total of 71 players spent at least part of it in the Swedish Hockey League before making their way back to the NHL the following season, with a lofty translation factor of 0.85.

Having become more of a development league since then, Sweden was not as popular a destination during the 2013 lockout. Only a dozen players made the move to the NHL from Sweden in 2013–14, which is only one more than the number of players who made the jump in 2014–15. However, just like in 2005, the translation factor remained a lofty 0.88, thanks to established NHL players like Cody Franson, Alexander Steen, Elias Lindholm, and Matt Duchene.

Over the past nine seasons (not counting the 2013 lockout), 89 players have made the move. Of those who played at least 20 games in both leagues, the average translation factor has dropped to 0.62.

PLAYERS WHO WENT FROM THE SWEDISH HOCKEY LEAGUE TO THE NHL, 2005-06 TO 2016-17

AGE	PLAYER	SEASON	GP	G	A	PTS	TEAM	GP	G	A	PTS
19	Nicklas Backstrom	2006–07	45	12	28	40	WSH	82	14	55	69
18	Anze Kopitar	2005–06	47	8	12	20	LAK	72	20	41	61
21	John Klingberg	2013–14	50	11	17	28	DAL	65	11	29	40
22	Tobias Enstrom	2006–07	55	7	21	28	ATL	82	5	33	38
18	Magnus Paajarvi	2009–10	49	12	17	29	EDM	80	15	19	34
22	Fabian Brunnstrom	2007–08	54	9	28	37	DAL	55	17	12	29
22	Mattias Janmark	2014–15	55	13	23	36	DAL	73	15	14	29
20	Artturi Lehkonen	2015–16	49	16	17	33	MTL	73	18	10	28
19	Marcus Johansson	2009–10	42	10	10	20	WSH	69	13	14	27
18	Erik Karlsson	2008–09	45	5	5	10	OTT	60	5	21	26
20	Marcus Kruger	2010–11	52	6	29	35	CHI	71	9	17	26
23	Melker Karlsson	2013–14	48	9	16	25	SJS	53	13	11	24
22	Mats Zuccarello	2009–10	55	23	41	64	NYR	42	6	17	23
19	Mattias Tedenby	2009–10	44	12	7	19	NJD	58	8	14	22
22	Sean Bergenheim	2006–07	36	16	17	33	NYI	78	10	12	22
19	Mika Zibanejad	2011–12	26	5	8	13	OTT	42	7	13	20
19	Alexander Wennberg	2013–14	50	16	5	21	CBJ	68	4	16	20
18	Victor Hedman	2008–09	43	7	14	21	TBL	74	4	16	20
21	Jakob Silfverberg	2011–12	49	24	30	54	OTT	48	10	9	19
18	Adam Larsson	2010–11	37	1	8	9	NJD	65	2	16	18
22	Patrick Thoresen	2005–06	50	17	19	36	EDM	68	4	12	16
22	Carl Gunnarsson	2008–09	53	6	10	16	TOR	43	3	12	15
22	Tom Wandell	2008–09	51	15	26	41	DAL	50	5	10	15
22	Erik Gustafsson	2014–15	55	4	25	29	CHI	41	0	14	14
33	Magnus Johansson	2006–07	52	8	28	36	CHI/FLA	45	0	14	14
18	William Nylander	2014–15	21	8	12	20	TOR	22	6	7	13

AGE	PLAYER	SEASON	GP	G	A	PTS	TEAM	GP	G	A	PTS
29	Pierre-Édouard Bellemare	2013–14	52	20	15	35	PHI	81	6	6	12
18	Jonas Brodin	2011–12	49	0	8	8	MIN	45	2	9	11
27	Ossi Vaananen	2007–08	45	7	8	15	PHI/ VAN	49	1	10	11
24	Johnny Oduya	2005–06	47	8	11	19	NJD	76	2	9	11
18	Jacob Josefson	2009–10	43	8	12	20	NJD	28	3	7	10
30	Joakim Lindstrom	2013–14	55	23	40	63	STL/ TOR	53	4	6	10
29	Josef Melichar	2007–08	50	0	8	8	CAR/ TBL	39	0	9	9
20	Anton Stralman	2006–07	53	10	11	21	TOR	50	3	6	9
30	Andreas Karlsson	2005–06	50	26	29	55	TBL	53	3	6	9
28	Rickard Wallin	2008–09	55	18	27	45	TOR	60	2	7	9
21	Patric Hornqvist	2007–08	53	18	12	30	NSH	28	2	5	7
20	David Rundblad	2010–11	55	11	39	50	OTT/ PHX	30	1	6	7
27	Jonas Frogren	2007–08	47	0	1	1	TOR	41	1	6	7
18	Jacob De La Rose	2013–14	49	7	6	13	MTL	33	4	2	6
23	Joel Lundqvist	2005–06	49	10	22	32	DAL	36	3	3	6
22	Jonas Holos	2009–10	51	1	13	14	COL	39	0	6	6
19	Anton Lander	2010–11	49	11	15	26	EDM	45	2	4	6
21	William Karlsson	2013–14	55	15	22	37	ANA/ CBJ	21	3	2	5
19	Gustav Forsling	2015–16	48	6	15	21	CHI	38	2	3	5
27	Mika Pyorala	2008–09	55	21	22	43	PHI	36	2	2	4

Minimum 20 games played in both leagues, excluding the 2012–13 season.

Given that you have to scroll all the way down to depth NHLers like Magnus Johansson and Pierre-Édouard Bellemare to find anyone over age 23, it isn't outrageous to describe Sweden as more of a developmental league. Half of the top-20 players on this list were teenagers when they arrived in the NHL.

That's why age has to be taken very seriously when translating player data from Sweden. The average translation factor for the nine 18-year-olds who went to the NHL for their age-19 season is a whopping 1.09. Half of these players actually scored at a *higher* rate in the NHL, with one more near miss.

These findings certainly make intuitive sense, since only truly talented players like Anze Kopitar and Erik Karlsson are going to get NHL opportunities that early in their careers. Beyond those elite prospects, the translation factor drops to 0.81 at age 19 and then to 0.53 for players in their 20s (and beyond).

The next interesting observation is that there doesn't seem to be the same relationship between scoring rates in the SHL and NHL that exists when players from other leagues move up to the NHL. While a very slight trend might become visible if you really squint, overall it looks more like Pac-Man got the stomach flu.

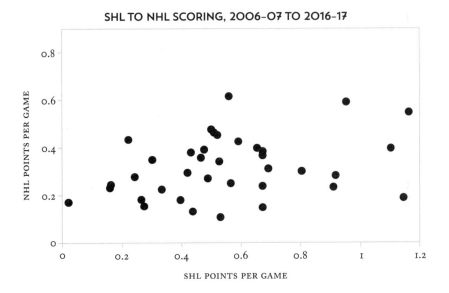

SHL TO NHL SCORING, 2006–07 TO 2016–17

As the chart illustrates, players who score really well in the SHL are only slightly more likely to do well in the NHL. Flip back to the table of 46 players, and consider those who dominated the SHL offensively (on the left) only to become nothing more than depth-liners in the NHL (on the right).

It's a good thing that the SHL is one of the few leagues with a website that offers additional statistics, like ice time, shots and scoring in various manpower situations, primary versus secondary

assists, and even faceoffs broken down by zone. Using NHL translations to help project a player's NHL scoring from Swedish league
data would otherwise be a total crapshoot.

Finland SM-liiga

Patrik Laine's and Sebastian Aho's immediate NHL success in
2016–17 came as quite a surprise to anyone who had been studying
the top Finnish league over the past decade.

Before Joonas Donskoi scored 36 points for the San Jose Sharks
in 2015–16, the last player to hit double digits in the NHL after
arriving directly from Finland's elite league was Cory Murphy, who
scored 17 points for Florida in 2007–08. That was the best total since
the 2005–06 lockout and tied with Tony Salmelainen's 17 points with
Chicago the year before.

Long ago, SM-liiga was one of the stronger European leagues.
From the moment the old amateur league SM-sarja went professional
in 1975, and players like Matti Hagman first arrived in the NHL in
1976–77, 268 players have gone directly from Finland to the NHL, with
an average translation factor of 0.46 (among the 99 players to play at
least 20 games in each league). That's about the same as the AHL.

SM-liiga has also been a popular destination for NHLers during
the various lockouts, which has further boosted its reputation.
Specifically, 20 players spent part of the 1994–95 lockout in Finland,
returning with a 0.58 translation factor. In 2004–05, that total jumped
to 33 players, with a translation factor of 0.66. Finally, 34 players called
SM-liiga their home during the 2012–13 lockout, and they came back
with a translation factor of 0.56. Finland can be a competitive place to
play hockey—when the NHL isn't in action, at least.

If you ignore the three lockout seasons, SM-liiga no longer
appears to deserve its strong historic reputation, prior to Laine and
Aho. Since the 2004–05 lockout, and ignoring the 2012–13 lockout,
49 players made the move before Laine, or just over five per season.
Among the 17 of 22 players in the following table who played at least
20 games in each league prior to Laine in 2016–17, many of whom

failed to remain in the NHL for more than a single season, SM-liiga had an average translation factor of 0.32. That places it below the other European leagues, below most U.S. college divisions, and on a par with the OHL. Puzzlingly, Finland was somehow still ranked third in world rankings by the International Ice Hockey Federation (IIHF).[39]

Now, thanks to Laine, Aho, and even Donskoi, SM-liiga is back on the scene with a mighty translation factor of 0.46, which places it on a par with the AHL, where it stood decades ago. They're back!

PLAYERS WHO WENT FROM THE FINNISH ELITE LEAGUE TO THE NHL, 2005–06 TO 2016–17

AGE	PLAYER	SEASON	GP	G	A	PTS	TEAM	GP	G	A	PTS
17	Patrik Laine	2015–16	46	17	16	33	WPG	73	36	28	64
18	Sebastian Aho	2015–16	45	20	25	45	CAR	82	24	25	49
22	Joonas Donskoi	2014–15	58	19	30	49	SJS	76	11	25	36
29	Cory Murphy	2006–07	45	13	37	50	FLA	47	2	15	17
24	Tony Salmelainen	2005–06	53	27	28	55	CHI	57	6	11	17
28	Petr Tenkrat	2005–06	36	10	21	31	BOS	64	9	5	14
24	Lasse Kukkonen	2005–06	56	11	16	27	CHI/ PHI	74	5	9	14
19	Teuvo Teravainen	2013–14	49	9	35	44	CHI	34	4	5	9
26	Lennart Petrell	2010–11	56	13	22	35	EDM	60	4	5	9
20	Mikael Granlund	2011–12	45	20	31	51	MIN	27	2	6	8
17	Jesse Puljujarvi	2015–16	50	13	15	28	EDM	28	1	7	8
25	Petteri Nokelainen	2010–11	46	11	16	27	PHX/ MTL	56	3	4	7
21	Markus Nutivaara	2015–16	50	6	16	22	CBJ	66	2	5	7
23	Anssi Salmela	2007–08	56	16	16	32	NJD/ ATL	26	1	5	6
26	Tommi Santala	2005–06	43	9	23	32	VAN	30	1	5	6
27	Jakub Nakladal	2014–15	50	3	12	15	CGY	27	2	3	5
26	Joonas Kemppainen	2014–15	59	11	21	32	BOS	44	2	3	5
26	Yohann Auvitu	2015–16	48	6	15	21	NJD	25	2	2	4
27	Ilkka Pikkarainen	2008–09	54	24	13	37	NJD	31	1	3	4

39. "2016 Men's World Ranking," International Ice Hockey Federation, accessed March 4, 2017, http://www.iihf.com/home-of-hockey/championships/world-ranking/mens-world-ranking/2016-ranking/.

AGE	PLAYER	SEASON	GP	G	A	PTS	TEAM	GP	G	A	PTS
20	Petteri Lindbohm	2013–14	37	1	5	6	STL	23	2	1	3
29	Stéphane Veilleux	2010–11	25	1	6	7	NJD/MIN	22	0	2	2
24	Dwight Helminen	2007–08	52	20	25	45	CAR	23	1	1	2

Minimum 20 games played in both leagues, excluding the 2012–13 season.

The SM-liiga is a good example of why it's important to keep translation factors up to date. Those who used Desjardins's original estimate of 0.49 were grossly inflating the results for years and are only now starting to get accurate translations.

Despite the fluctuations, SM-liiga translations are fairly reliable: the trend line is clear and relatively gradual. Those who score at a higher rate in Finland enjoy more success in the NHL, without much variation and with only three notable outliers.

FINLAND SM-LIIGA TO NHL SCORING, 2006-07 TO 2016-17

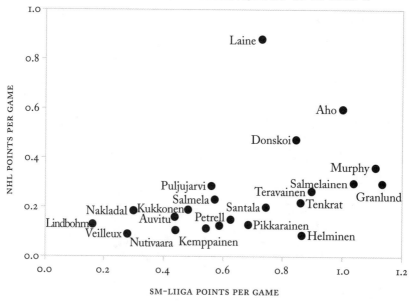

These three exceptions are what makes studying SM-liiga so fascinating. As can be seen on the horizontal axis, eight players outscored Laine while in Finland, including fellow teenagers Aho and

Teravainen. And yet, Teravainen fell right on with expectations, with just 9 points in 34 games the next year in the NHL, and even Aho's impressive scoring totals were nowhere near Laine's. Either Laine is a particularly exceptional player or got incredibly lucky, or age has a particularly dramatic impact in Finnish translations.

To help sort things out, SM-liiga has recently introduced many of the same advanced metrics as the SHL, making it possible to integrate factors such as ice time and power-play scoring into the translation factor. All that work never seemed worthwhile when Finland was only producing 17-point players, but with the arrival of players like Donskoi, Aho, and Laine, there may be more interest in further research.

Switzerland NLA

To some, it didn't make sense that an elite prospect like Auston Matthews would choose to play in Switzerland instead of the CHL or NCAA or even one of the stronger European leagues. However, there was evidence that showed Switzerland's National League A was on a par with those other options.

Even with Matthews included, the NLA appears to produce the most predictable NHL projections of all European leagues. Consider the clear pattern on the following chart, with a smooth upward trend from which very few (other) players significantly deviate.

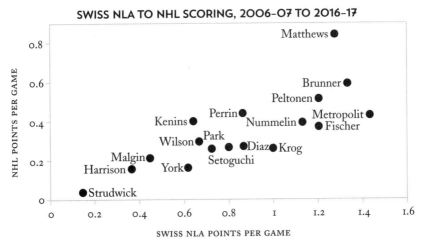

SWISS NLA TO NHL SCORING, 2006-07 TO 2016-17

How Can We Compare a Player's Stats Between Leagues?

Matthews helped shine a new light on a league that was often overlooked as the source of some of Europe's most competitive hockey. The Swiss NLA is actually one of the continent's oldest leagues, going back over 100 years if you include earlier incarnations.

Starting with Mark Pavelich in 1981–82, 45 players made the jump before the 2005 NHL lockout, averaging a 0.46 translation factor. Then 40 players who spent at least part of the lockout in Switzerland returned to the NHL for 2005–06, averaging a translation factor of 0.67. Another 29 players made the jump during the 2012–13 lockout, with an average translation factor of 0.64.

Since then, 34 players have made the move to the NHL, not including the 2012–13 lockout, with an average translation factor of 0.43 among the 19 who played at least 20 games in both leagues. That makes Switzerland's NLA roughly on a par with Finland's SM-liiga as the third-strongest European league, which is on a par with the best U.S. college hockey and only a slight step back from the AHL. And so, it actually made perfect sense for Auston Mathews to play his pre-draft season in Switzerland, and others may soon follow.

Prior to Matthews, there were no players who had a major impact when coming directly from the NLA, probably because 11 of the preceding 15 players were already in their 30s. That may also explain why their scoring has been so predictable. For whatever reason, NHL teams have decided that only veterans can make the direct trip and have otherwise required that the league's talent pass through the AHL.

PLAYERS WHO WENT FROM THE SWISS NLA TO THE NHL, 2005-06 TO 2016-17

AGE	PLAYER	SEASON	GP	G	A	PTS	TEAM	GP	G	A	PTS
18	Auston Matthews	2015–16	36	24	22	46	TOR	82	40	29	69
32	Ville Peltonen	2005–06	39	22	25	47	FLA	72	17	20	37
30	Éric Perrin	2005–06	44	13	25	38	TBL	82	13	23	36
31	Glen Metropolit	2005–06	44	24	39	63	ATL/STL	77	14	19	33
26	Damien Brunner	2011–12	45	24	36	60	DET	44	12	14	26
28	Tom Pyatt	2015–16	42	11	18	29	OTT	82	9	14	23
33	Petteri Nummelin	2005–06	38	13	30	43	MIN	51	3	17	20

AGE	PLAYER	SEASON	GP	G	A	PTS	TEAM	GP	G	A	PTS
25	Raphael Diaz	2010–11	45	12	27	39	MTL	59	3	13	16
32	Dick Tarnstrom	2006–07	44	3	26	29	EDM/CBJ	48	3	11	14
34	Richard Park	2010–11	47	15	19	34	PIT	54	7	7	14
23	Ronalds Kenins	2013–14	39	8	17	25	VAN	30	4	8	12
29	Devin Setoguchi	2015–16	30	11	13	24	LAK	45	4	8	12
30	Patrick Fischer	2005–06	44	21	32	53	PHX	27	4	6	10
19	Denis Malgin	2015–16	38	5	12	17	FLA	47	6	4	10
32	Landon Wilson	2007–08	30	13	7	20	DAL	27	2	6	8
35	Jason York	2005–06	34	3	18	21	BOS	49	1	7	8
30	Jason Krog	2005–06	29	5	14	29	ATL/NYR	23	3	3	6
26	Jay Harrison	2008–09	41	6	9	15	TOR	38	1	5	6
31	Jason Strudwick	2006–07	34	2	3	5	NYR	52	1	1	2

Minimum 20 games played in both leagues, excluding the 2012–13 season.

Oddly, very few of those coming over are Swiss, just Raphael Diaz, Damien Brunner, and Denis Malgin (though Mark Streit came over in an earlier time frame). Other Swiss players like Luca Sbisa, Sven Baertschi, Yannick Weber, and Nino Niederreiter came in through the Canadian junior leagues, and Roman Josi was brought in through the AHL.

It would be ideal to close this section with a closer look at Mathews, who played his 18-year-old season with Zurich instead of in the CHL or NCAA Division I. Unfortunately, Mathews is in uncharted territory, as 23-year-old Ronalds Kenins is the youngest player to compete in the Swiss NLA and go on to play even 20 games in the NHL the following season, and he's not exactly a good statistical comparison. Plus, Matthews's preceding two seasons were in the USHL, which has no formal translation factor because virtually nobody goes directly from that league to the NHL.

Using the league's prior 0.39 translation factor, Mathews's 44 points in 34 games works out to 41 points over a full 82-game NHL season. As such, it looked like he really crushed it by scoring 69 points instead. However, since 17-year-old superstars in the OHL

— D.O.P.S. WHEEL OF JUSTICE —

('MON UP, BRAD! GIVE IT A SPIN...

PLAYER SAFETY RULING

have a translation factor that's over 40% higher than the older players, Mathews's actual translation was closer to 60 points. (Auston Matthews was actually 18 years old in Switzerland, but with a late birthday.) He still outperformed expectations, but not by as dominant a margin as it originally appeared. In fact, now that the translation factor for this league has been increased to 0.43, it will help forecast future prospects even more accurately.

Other Leagues

Very few players jump to the NHL directly from any of the other European or North American leagues, which makes translation factors both difficult and pointless to compute.

For example, the Czech Extraliga used to be included in these types of studies, but since the 2004–05 NHL lockout there have been only three players to compete in at least 20 games in both leagues (in a non-lockout season), all of which occurred in 2007–08 (Jan Hlavac, Jaroslav Hlinka, and Vladimir Sobotka). Any attempt to calculate a translation factor on such limited and out-of-date information would be a waste of time and apply to very few current players, if any.

Those who insist on getting some kind of estimate for one of the less popular leagues can use the same two-step method that Desjardins used in his original study to calculate translation factors for the USHL, Tier II Junior, and the NAHL or in his later articles specific to the Canadian Junior A leagues and Minnesota high school hockey.[40] More recently, Kent Wilson used this two-step method to project the scoring of Mikael Backlund, from Sweden's Allsvenskan league.[41]

Essentially, when there isn't enough data for players who went directly from the target league to the NHL, we can calculate a translation factor to an intermediate league, such as the CHL in Desjardins's case or the SHL in Wilson's, which we can then multiply by the known NHL translation factor for that league.

To further explain this process using the latter example, Wilson gathered all of the data of players going from Allsvenskan to his chosen intermediary league for which an NHL translation factor was already known, the Swedish Elitserien, and he calculated a translation factor using the same process described in this chapter. He then multiplied the two numbers to arrive at an NHL translation factor of 0.36, which proved handy when Filip Forsberg and David Pastrnak came along.

When pioneering this process, Desjardins estimated NHL translation factors for U.S. high school hockey between 0.05 and 0.07, the Canadian Junior A leagues between 0.13 to 0.18, and the USHL and the NAHL at the top of that range, possibly a point or two higher. In theory, we can also use the same process to calculate translation factors for leagues like the Czech Extraliga and bolster the leagues for which we don't have a lot of data, like the SM-liiga and Swiss NLA.

Due to the work involved, I'm not aware of anyone who has gone to the trouble of publicly updating any of these secondary league translation factors in the past four or five years. While that's no great loss for fans and pundits, who can simply wait until a player

40. Gabriel Desjardins, "Projecting Junior Hockey Players and Translating Performance to the NHL," Behind the Net, accessed DATE, http://www.behindthenet.ca/projecting_to_nhl.php; Gabriel Desjardins, "Canadian Junior A/NCAA/Major Junior Equivalencies," *Behind the Net* (blog), September 8, 2008, http://www.behindthenet.ca/blog/2008/09/canadian-junior-ncaamajor-junior.html; Gabriel Desjardins, "NHL Equivalences, Minnesota High School Hockey," *Hockey Prospectus* (blog), June 23, 2009, http://www.hockeyprospectus.com/puck/article.php?articleid=192.

41. Kent Wilson, "Allsvenskan NHL Equivalence," Flames Nation, updated November 5, 2012, http://flamesnation.ca/2012/11/5/allsvenskan-nhl-equilvance.

competes in one of the major leagues covered in this chapter, NHL front offices definitely require a complete set of translation factors, for the NHL Entry Draft. For example, in the decade following the 2005 lockout, 25 USHL players were selected in the first round, eight were selected from the American high school system, six were selected from the Allsvenskan, and at least one player was selected from each of 10 other leagues. For the NHL, the investment required to put together career projections is more than worthwhile. Our work, however, is complete.

Closing Thoughts

There are seven key findings in this chapter that are worth repeating:

1. As we have seen with players like Panarin and McDavid, NHL league translations are a surprisingly accurate way to project a rookie's NHL scoring, and they provide enough data to map out a player's career scoring arc.
2. For players aged 19 or younger, it is absolutely critical to consider age in their projections, right down to the birth month. At the very least, it's important to stay on top of which players have late birthdays.
3. Prospects who outperformed their calculated NHLe are at great risk of a scoring drop in their second year in the NHL and beyond.
4. It's important to consider the quality of the league in which a prospect played, and even the specific conference in the case of the NCAA. With the exception of the KHL (which is on the decline), the European leagues also qualify as developmental leagues.
5. For outlying players, rather than use the traditional method of a straight translation factor anchored at zero points per game, assuming a base level of scoring and translating the player's scoring at a reduced slope is a more accurate method.
6. Since this issue with outlying players is due to the reduced ice

time a player received when coming to the NHL, especially with the man advantage, we can also achieve a more precise calculation by taking such factors directly into account.

7. Unfortunately, the lack of available statistics from all but the two main Nordic leagues makes such studies challenging, if not impossible.

Let's close with the translation factors once again, this time broken down by the adjustment that's required for the NHL's higher or lower goals-per-game scoring levels and then adjusting by the remaining component that's sometimes referred to as the league's quality.

EXPANDED NHL TRANSLATION FACTORS, AS OF 2016–17

LEAGUE	SCORING	QUALITY	FACTOR
Kontinental Hockey League (KHL)	1.08	0.69	0.77
Swedish Hockey League (SHL)	1.07	0.54	0.62
American Hockey League (AHL)	0.97	0.48	0.47
Finland SM-liiga	1.07	0.41	0.46
Western Collegiate Hockey Association (WCHA, pre-2013)	–	–	0.44
National Collegiate Hockey Association (NCHC)	–	–	0.43
Switzerland (NLA)	0.95	0.45	0.43
Hockey East	–	–	0.38
Big 10	–	–	0.33
Central Collegiate Hockey Association (CCHA, now defunct)	–	–	0.32
Ontario Hockey League (OHL)	0.80	0.38	0.31
Western Hockey League (WHL)	0.87	0.33	0.28
Quebec Major Junior Hockey League (QMJHL)	0.81	0.30	0.25
ECAC	–	–	0.23

CAN A GOALIE'S STATS BE COMPARED BETWEEN LEAGUES?

In 2014–15, AHL rookie Matt Murray set the league record with a shutout streak that lasted 304 minutes and 11 seconds, set the single-season rookie record with 12 shutouts, was named to the First All-Star Team and the All-Rookie Team, won the Baz Bastien Memorial Award for best goalie, and won the Red Garrett Memorial Award for best rookie.

Given that level of success, his career 0.925 NHL save percentage isn't entirely unexpected, nor is it a complete surprise that he played a key role in Pittsburgh's Stanley Cup victories in 2015–16 and 2016–17 or that Pittsburgh exposed their long-time number one goalie Marc-André Fleury in the expansion draft instead of him. However, it draws a great deal of attention to the question of whether or not a goalie's AHL data can be used to project his NHL career—and help find the next Matt Murray.

Statistically, the problem with goaltending data has always been the volatility of save percentage. It can rise and fall almost unpredictably from one season to the next, even when a goalie remains within the same league and on the same team. In this case, how can the Penguins be confident that Murray's 0.925 save percentage in 62 NHL games is not just a temporary hot streak? After all, Ottawa's

Andrew Hammond has a 0.923 save percentage in 55 NHL games, and nobody would give up Fleury for him.

The key to answering this question is to increase the amount of data being used to evaluate goalies by including their stats from other leagues. In this case, it's quite informative to note that Murray had a 0.936 save percentage in 72 AHL games, which is far better than Hammond's 0.903 save percentage in 80 AHL games. But how high does his save percentage need to be, and over how many games, before it starts to inform his expectations in the NHL?

The obvious first step is to see if there's any relationship between a goalie's save percentage in the AHL and in the NHL. To my knowledge, the first statistician to look at this publicly was Stephan Cooper back in 2012.[42] However, his results were not very encouraging, and this appears to have discouraged any further investigation.

Cooper's AHL-to-NHL translation methodology was the same as described in the previous chapter for skaters, except that he looked at the two seasons prior to and after heading to the NHL, not just one. In total, his study involved all 39 goalies who played at least 20 games total for four seasons in both leagues between 2000–01 and 2011–12.

His results? On the whole, he found that save percentages dropped from 0.917 in the AHL to 0.910 in the NHL. But on an individual basis, Cooper found that "the relationship between the AHL performance and the NHL performance proved to be pretty much random" and concluded that "AHL single season results have little bearing on goaltender (save percentage) in the NHL the following season."

I'm usually not one to waste my time recreating other people's work, especially when they hit a dead end, but in the quest to improve our understanding of goaltenders by expanding the pool of data, I recreated his study but with a broader net. By going all the way back to 1993–94, which was the first season that the AHL officially recorded save percentage, it was possible to build a data set of 113 goalies who played at least 20 games in the AHL before

42. Stephan Cooper, "AHL to NHL Translations: Save Percentage," *Eyes on the Prize* (blog), July 11, 2012, https://www.habseyesontheprize.com/2012/7/11/3147551/ahl-to-nhl-translations-save-percentage.

playing at least 20 in the NHL the following season.[43] To maximize the number of data points, I dropped the two-season requirement to just a single season.

On an individual basis, my results were similar to Cooper's. That is, there was very little correlation between how goalies performed in the AHL relative to their following season in NHL. As a graphical demonstration, consider the data points in the following chart. To me, it just looks like the spread of pellets from a shotgun, without any obvious relationship or trend line.

AHL TO NHL SAVE PERCENTAGES, 1993-94 TO 2014-15

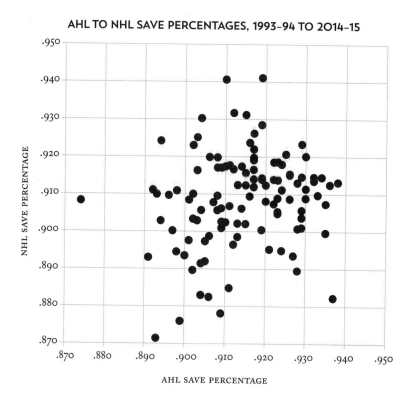

The problem could be that 20 games is an awfully small sample size. Even a goalie who played 20 games in consecutive NHL seasons

---

43. AHL goaltending data from the American Hockey League, accessed September 8, 2017, http://www.theahl.com.

could have a save percentage of 0.900 one season and 0.920 the next. That doesn't mean that his stats have no predictive value, but rather that there's not enough of it.

When there's too much individual variation, trends and patterns can sometimes be revealed by grouping data. It's a very common and useful technique. In this case, when the AHL goalies are evenly divided into the five groups featured in the chart below, some patterns do begin to reveal themselves.

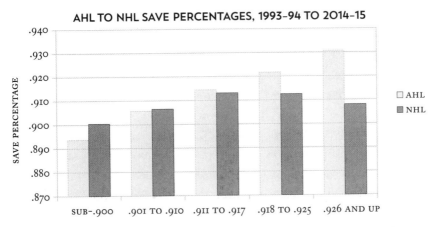

AHL TO NHL SAVE PERCENTAGES, 1993–94 TO 2014–15

From this new perspective, a goalie's NHL save percentage does not appear to be completely unrelated to his AHL performance. For example, the group of goalies with AHL save percentages between 0.911 and 0.917 were demonstrably better in the following NHL season than those in the 0.901 to 0.910 group. If anything, both groups largely retain their save percentages in the NHL, with a slight tendency toward the average.

The fringe cases are the more interesting results, as further increases in AHL save percentages beyond 0.917 actually result in *lower* NHL save percentages. That seems like an odd result at first glance, but it makes sense when you consider selection bias. That is, the types of goalies contained in each of these categories were not randomly selected and are not the same. Qualifying for certain groups means possessing certain qualities that will influence the results. That's classic selection bias.

In the case of the group with AHL save percentages of 0.926

and up that went on to the NHL, they are far more likely than the other groups to be goalies who simply got hot and lucky. If so, that would explain why the average NHL save percentage for that group is actually lower than the groups with more average AHL save percentages. In essence, some goalies performed so well in the AHL that they were perhaps promoted to the NHL before they were truly ready.

Consider Jacob Markstrom. At age 21, he recorded a sparkling 0.928 save percentage in 32 games for the San Antonio Rampage. He was consequently brought up by the Florida Panthers and posted just a 0.901 save percentage as a 22-year-old the following season. I'm not sure he would have received that early NHL opportunity if his AHL save percentage had been 0.914. That's the type of goalie who's overrepresented in that last group, and that could be why the group's NHL save percentage is lower.

On the flip side, consider those who performed poorly in the AHL but who somehow secured an NHL opportunity anyway. These are obviously not your typical goalies either, but a mix of goalies who had previously proven themselves, desperately needed injury replacements, or young and developing goalies in whom the organization believed so strongly that they were brought along even without AHL success. The end result is a low NHL save percentage, but not as low as expected, given the AHL data.

In the end, what can be done with these results? It's tempting to simply set AHL goaltending data aside as having too little predictive value, but data shouldn't be ignored unless it's completely irrelevant, like jersey numbers and hot dog sales. As such, AHL data should be included, especially since the third group was clearly more effective than the second.

To find the middle ground between the shotgun chart and the five big groups, the following chart was created by taking every single goalie and adding the results of the four goalies with the most similar AHL save percentages in either direction. Therefore, each point on the following chart represents nine goalies, which is enough to start painting a useful picture.

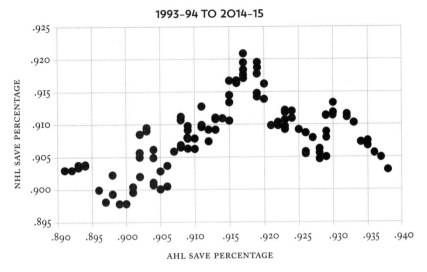

There is a usable relationship between AHL and NHL save percentages, but it's not a straight line or a standard curve. To the eye, there appears to be a good trend line from about the 0.905 save percentage to somewhere short of 0.920. Then it actually descends, possibly for the reasons mentioned previously, with the Markstrom example. Curiously, there also appear to be spikes at the 0.903, 0.917, and 0.930 marks—or every 14 points or so. That could simply be the impact of a handful of goalies skewing a relatively modestly sized data set, but we can't be certain.

Based on this, our system should actually project a lower NHL save percentage for an AHLer with a 0.928 save percentage than someone who posts a 0.917—much lower! That may sound crazy, but the argument is that a 0.917 goalie who fights his way onto the NHL roster is likely to be a legitimate 0.917 goalie, whereas a 0.928 goalie who gets a shot in the NHL is more likely to be a goalie who got hot and whose luck is about to run out.

Of course, there are ways to figure out which goalies simply got hot: more data. Thus far, we have only been studying single-year translations, and *translations are not projections*. If we use multiple seasons' worth of AHL data, especially with the statistical regression and age curves described in the earlier chapter on the most valuable

— OUTDOOR GAME SCHEDULE CONFLICT —

goalies, then hot and cold streaks would be averaged out over time, and very few goalies would remain outside the 0.905 to 0.917 range.

Consider Ben Bishop, who posted a 0.930 save percentage in the AHL followed by 0.920 in the NHL. A simple 4-2-1 weighting of his final three AHL seasons results in a much closer 0.921 projection. Since that's just one of about a dozen similar examples, the obvious but key takeaway is to use several seasons when using a translation factor in statistical projections.

Incidentally, that approach doesn't work when projecting Markstrom, who remains an unusual case. Over his entire career, Markstrom has a save percentage of 0.921 in 163 AHL games but just 0.906 in 109 NHL games. Most hot AHL goalies land on a more modest save percentage in the long run, which matches how they perform in the NHL, but that doesn't happen every single time. Then again, Markstrom is only 27 years old, and time has yet to render its verdict.

That covers the AHL, but what about other leagues? Of the 60 goalies who began the 2015–16 season in the NHL, between 46 and 51 came directly from the AHL, depending on which standard you use, while no more than three goalies came directly from any other single league. (As an interesting side note, the age at which goalies make the jump is, for the most part, evenly divided from ages 22 through 26, with a handful of cases in the two years on either end.) While a 19-year-old skater can find himself in the NHL right away, no matter where he came from, goalies generally aren't used until they have been seasoned in the AHL for years and are already in their prime. As such, translating goaltending data from other leagues to the NHL is more about translating it to the AHL first, but that's a topic for another day.

Closing Thoughts

While Cooper's initial misgivings are perfectly understandable, it appears that we can use AHL goaltending data as part of an NHL projection. As for other leagues, there's not enough data to forecast their NHL future, since about 75% to 85% of goalies go through the AHL first. The next step appears to be to build an AHL translation model.

Even with the AHL data, extra effort is required to correctly forecast the future of those with particularly high or low save percentages, while bearing in mind that translations are not projections themselves but rather a piece that can form a part of a projection.

As described in the earlier chapter on the most valuable goaltender (and in potentially excessive detail in *Stat Shot*[44]), we build a projection with several seasons' worth of data, use statistical regression to remove random variation, and apply a proper aging curve. Translated data is just meant to serve as one (or more) of those seasons, and it must still go through that process in order to produce a meaningful projection.

So who is the next Matt Murray? At the time of writing, the best bet might be Juuse Saros of the Nashville Predators. He's only 22 years old, had a 0.929 save percentage in 91 games in SM-liiga, followed by a 0.924 save percentage in 53 games with the Milwaukee Admirals of the AHL, and already has a 0.922 save percentage in 22 games in the NHL. The only problem is that he is 5-foot-11. If he succeeds, it could pave the way for many others.

44. Rob Vollman, "What's the Best Way to Build a Team?," *Stat Shot*, 12–74.

HOW CAN STATS BE PLACED IN CONTEXT?

"This shows how stupid the Corsi thing is . . . here's a guy that every coach would want . . . this is how stupid guys come up with . . . trying to earn a living . . . they come up with dumb . . . they said this guy is the worst hockey player in the league . . . this guy is unbelievable and they call him the worst from that dumb thing that you did . . . and he's the worst player? I'll tell you one thing . . . Vancouver loves this guy . . . he is unbelievable . . . and that dumb-dumb system you're talking about . . . I would love to have Ryan Johnson on my team and every coach would have him on too . . ."

—Don Cherry, *Hockey Night in Canada*, March 27, 2010[45]

The beloved Don Cherry was making one of the classic mistakes in the world of hockey analytics, by failing to take context into account. (The other mistake was believing that someone could earn a living on Corsi, especially back in 2010.) Yes Johnson had the worst Corsi in the NHL, but that doesn't mean he should have been considered the worst player.

Regardless of the stat, we must always take context into account. In this case, Johnson was being used primarily in the defensive zone and with relatively unskilled linemates like Tanner Glass, Rick

45. Gabriel Desjardins, "This Shows How Stupid the Corsi Thing Is . . .," *Arctic Ice Hockey* (blog), March 28, 2010, https://www.arcticicehockey.com/2010/3/28/1394052/this-shows-how-stupid-the-corsi.

Rypien, and Darcy Hordichuk. In terms of usage, Johnson was the opposite of Warren Young.[46] It's obviously not a good sign that Johnson's Corsi was so bad, and indeed, his NHL career lasted only 34 more games, but you can't just take a single stat and run all the way to the end zone with it, like Cherry did.

Even in these very pages, we have stumbled on the issue of focusing too much on a single stat. While tackling how to translate an individual's stats from another league into an NHL standard, the importance of determining a player's role became clear. After all, there's a big difference between the NHL fate of a player who scored 50 points as a scoring-line forward on the powerhouse St. Petersburg team and one who did so as a checking-line centre on a basement-dwelling team in Finland.

There are a lot of different ways to put stats in context, and we'll be focusing on player usage charts, which were introduced for just this purpose back in the summer of 2011.[47] Here's how it looks for Johnson and the 2009–10 Canucks, for example. Johnson is way on the left side of the chart and has the big white circle that indicates a poor Corsi.

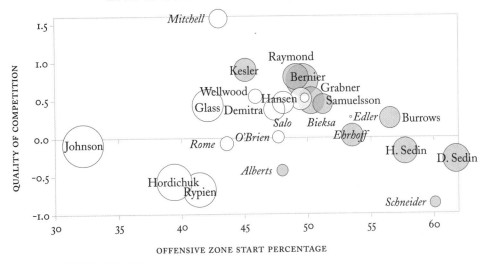

2009–10 VANCOUVER CANUCKS, PLAYER USAGE

46. At age 29, Young broke into the NHL alongside rookie Mario Lemieux (who I hear was a pretty good hockey player). Young scored 40 goals on 130 shots. Separated from Lemieux the next season, he scored 30 goals in his remaining 136-game career.

47. Rob Vollman, "Winnipeg OZQoC Charts," *Arctic Ice Hockey* (blog), June 20, 2011, https://www .arcticicehockey.com/2011/6/20/2233834/winnipeg-ozqoc-graphs.

So what are all of these numbers and circles? One of our field's most noteworthy developments, player usage charts are a graphical representation of three modern statistics that describe in what situations players are used.

1. Offensive zone start percentage is on the horizontal axis. It is based on what percentage of a player's shifts start in the offensive zone (ignoring shifts that start in the neutral zone or on the fly). Notice that Johnson is on the far left, since he got buried in the defensive zone.

2. Quality of competition is on the vertical axis. It is the average Corsi of a player's opponents weighted by how much time they faced each other. Sometimes, it's the weighted average ice time of a player's opponents. Instead of opponents, quality of teammates is sometimes used, depending on what aspect of the game is being studied. Here, Johnson is on the horizon, since he faced roughly average competition.

3. Relative Corsi, which is a player's shot-based plus/minus per minute relative to his teammates, is used for the size and shading of the circles around each player's name. Big shaded circles indicate the player is very good, and the big white circle around Johnson, and his frequent linemates like Glass, Hordichuk, and Rypien, indicates something quite bad. Some of the more advanced versions of these charts integrate a player's average ice time into these circles as well.

Put it all together, and player usage charts quickly provide a perspective of how a team is deploying its players and how effective that usage has been. It allows you to effortlessly avoid Cherry's mistake, and it puts both statistical and subjective information in context. It also allows you to compare apples to apples by finding other players around the league in the same role as Johnson or, better yet, more interesting players. Notice how Daniel and Henrik Sedin were used primarily in the offensive zone and against secondary opponents, while Ryan Kesler took on the league's top opponents in both zones. On the blue line, notice the difference between Mathieu Schneider and Willie Mitchell. This is exactly the kind of information that can help us place each player's contributions into the right context.

Given their usefulness, player usage charts quickly spread through the analytics community, helped kick-start an era in which the context in which players are used is more consistently and accurately taken into account, and inspired a new trend—our community's newest statisticians aim to express their innovations visually, rather than with tables of numbers.

If there's already a solution for placing a player's stats in context, why are there still so many pages left in this chapter? Well, player usage charts aren't exactly perfect. Then again, what statistical perspective ever is or was meant to be? As always, we do our best with the available resources and make improvements as new advances are introduced. There have been some new developments that can be incorporated into these charts to improve the way in which we determine a player's usage.

The even greater issue with player usage charts is that they use data that is only available in the NHL and for less than a decade. We'll start by crafting a simplified version of player usage charts that can help establish the context in leagues and eras beyond the modern-day NHL.

Simplified Player Usage Charts

Fundamentally, player usage charts are intended to be a visual representation of how players are being deployed, whether it's in an offensive or defensive fashion and whether it's with and/or against top lines or depth lines. The exact statistics that are used to establish this usage is a secondary concern.

Typically, the vertical axis represents whether players are lining up with and/or against top competition, the depth lines, or somewhere in between. Sadly, these quality of competition metrics are only available on independent sites like Behind the Net, and occasionally they pop up temporarily whenever a hobbyist wants a job in the NHL.[48] Generally speaking, these metrics are not available outside the NHL nor beyond the recent past, nor are they something that fans can easily calculate for themselves.

For the simplified version of the charts, we'll need a statistic that serves roughly the same purpose but is far easier to find and/or calculate ourselves. How about average ice time? After all, those who get more ice time, especially in the third period or in close games, tend to be the same players who take on top opponents or play with top linemates. There are exceptions to this rule, but then again, there are exceptions to the rule with every statistic, including the existing quality of competition metrics.

Even-strength ice time may not be the perfect solution, since it's only available in the NHL back to the 1997–98 season and still isn't available in other leagues, beyond the two main Nordic leagues. Even a player's overall ice time is unavailable for most leagues, including the American Hockey League. There are methods that we used prior to 1997–98 to estimate a player's ice time, but they aren't simple. If this approach proves to be promising when used with the actual ice time data that we have, then we can pursue that option to look at other leagues and/or to go further back in history.

As for the horizontal axis, it is meant to indicate if the player is used primarily in an offensive or defensive fashion. In the modern-day NHL, we can estimate that using zone starts, or possibly with even newer innovations (which we'll explore later). Outside the NHL or for situations prior to 2006–07, when that information was first available in the NHL, we'll have to use something else.

As a child, I often looked at a player's shots per game as a crude estimate of whether a player was offensive minded or not. For example, I looked at the famous 1976–77 Montreal Canadiens and concluded that Steve Shutt, Guy Lafleur, and Jacques Lemaire must have been the top scoring line because they each averaged at least 3.6 shots per game, while no other forward was above 2.1. Similarly, I concluded Doug Jarvis must have been a defensive specialist because he played 80 games and took only 88 shots. On the blue line, Guy Lapointe led the team with 3.8 shots per game, Larry Robinson took

---

48. The Behind the Net website is www.behindthenet.ca, but it hasn't been updated since the 2015–16 season.

2.6, and everybody else was at 1.4 or below. That gave me a pretty good indication of who was focused on scoring and who wasn't.

Even today, I think that's a pretty reasonable way to figure out whether players are being used offensively or not. Daniel Sedin averaged 3.6 shots for the Canucks in 2009–10, and Johnson averaged 0.3. On the blue line, Christian Ehrhoff and Alexander Edler averaged 2.3 and 2.1, while Mitchell and Aaron Rome averaged 1.0.

Of course, that's not a perfect solution, since it ignores the contributions of playmakers like Henrik Sedin, who averaged 2.0 shots per game but was every bit as offensive minded as his brother Daniel. Plus, calculating a player's shots on a per-game basis makes it hard to tell the difference between players who aren't offensively minded and those who are but didn't get a lot of ice time.

That's why I settled on using each player's scoring rate on the horizontal axis instead. After all, the more points a player scored per minute, the more often he was probably playing in the offensive zone. At the very least, the more likely he *should* have been playing in the offensive zone. It's not a perfect solution but, then again, neither are shots per game or offensive zone starts.

That just leaves the third and final component of player usage charts: the shaded circles around each player. Getting right to the point of this entire chapter (in case you thought I had forgotten), the statistic being placed in context is Corsi. In the larger sense, it's meant to represent how well the team has performed with that individual on the ice, so it can actually represent *any* stat, including those that don't make Don Cherry lose his mind. Ideally, player usage charts place a statistic based on wins or goals in context, but there isn't enough of either to evaluate a player in single-season doses. That's why shot-based metrics like Corsi are used instead, since there are a lot more shot attempts per game (over 100) than goals (around 5.5) or wins (exactly 1).

In this case, the problem is that individual Corsi is generally available only in the NHL and only for the past decade. Where and when shot-based data is unavailable, there's really no choice but to use goal-based data, which means using plus/minus, warts and all. But before you scrunch up your face in disgust at the mere mention of such an allegedly flawed statistic, remember that a lot of the failings with plus/minus are actually addressed in a player usage chart:

- it will be presented on a per-minute basis, which will remove the bias that affects those with extra ice time;
- it is measured relative to a player's teammates, which will avoid the skewing effects of playing on a particularly good or bad team; and
- it is placed into the proper context of each player's usage.

So yes, I know that Lafleur and Shutt were +136 and +105 in 1976–77 while Jarvis and Gainey were +38 and +33, but that information will be placed in context on a player usage chart. Indeed, placing that kind of otherwise deceiving information into perspective is the entire point of a player usage chart, so we should take advantage of that.

Of course, there are still a few remaining problems with plus/minus, like the inclusion of short-handed goals, the skewing effect of playing with specific linemates, and the dramatic impact that shooting luck can have on small samples. While these issues are mostly unavoidable, they can be somewhat mitigated by adopting the standard Hockey Abstract philosophy of using three-year sample sizes to evaluate players. It also means that there will never be a statistical substitute for our own (subjective) knowledge and common sense.

Before dwelling on the anticipated problems that may or may not transpire, let's start by looking at a first cut of these simplified player usage charts and see if they serve as a reasonable starting point.

The 2010–11 Vancouver Canucks

Rather than using Johnson and the 2009–10 edition of the Vancouver Canucks, let's advance one season and use the more familiar example: the classic 2010–11 Vancouver Canucks that came within a single bounce of winning the Stanley Cup (and preventing a riot).

That particular season's player usage chart stands out because of the clearly established and extreme fashion that coach Alain Vigneault deployed his players. Also, instead of looking at fourth-line players like Johnson, it's more interesting to look at players like the Sedins, whose deployment change caused them to break out from solid, point-a-game players to Art Ross Trophy winners at ages 29 and 30.

Whether you used your eyes or relied on the figures, it was obvious that the Sedins were deployed offensively, that Ryan Kesler took on the top opponents, and that players like Manny Malhotra were buried in the defensive zone. (Oh, and the Canucks were bombed every time Glass was on the ice.) As such, the following chart makes sense even to those who are setting their eyes on a player usage chart for the very first time. (Well, for the second time, unless you had your eyes closed a few pages ago.)

2010-11 VANCOUVER CANUCKS, PLAYER USAGE

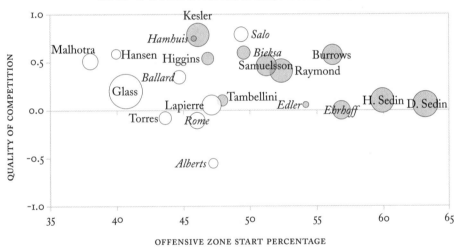

If we accept that this chart is a largely accurate representation of how the Canucks were deployed in 2010–11, then we can gauge the accuracy of the new simplified player usage charts by comparing them to this baseline. (However, in actuality, the player usage charts generally use a three-season weighted average, not just the 2010–11 season.)

The next chart takes all of the adjustments mentioned in the preceding section into account. Specifically, each player's scoring rate is on the horizontal axis, instead of offensive zone start percentage, and each player's average ice time is on the vertical axis, instead of quality of competition or teammates. To keep the defencemen on the same scale as the forwards, their scoring rates are multiplied by 2.5 and their ice time by 0.75.

Finally, the sized and shaded circles are based on each player's plus/minus per 60 minutes relative to expectations. In other words, I worked out what every player's expected plus/minus should have been, based on their team and their ice time, and then compared that to their actual figure. As in traditional player usage charts, shaded circles indicate those who had a higher plus/minus than expected, and the size of the circle represents the extent of that advantage (on a per-minute basis).

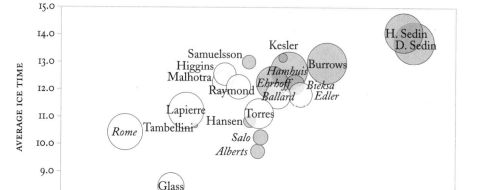

2010–11 VANCOUVER CANUCKS, SIMPLIFIED PLAYER USAGE

While players are scattered all over on a regular player usage chart, they appear to all be grouped along a relatively straight and inclined line going from the bottom left to the top right on the new simplified versions. This actually makes sense, given that players who score at a higher rate (horizontal axis) tend to get more ice time (vertical axis).

Just because they look different doesn't mean that these simplified player usage charts are flawed, just that they have to be read a little bit differently than the standard ones. Rather than judging players based on a horizontal line anchored at zero, we have to tilt our perspective and consider them relative to an imaginary line going from the bottom left to the top right.

For starters, we can figure out a team's top-to-bottom depth chart by reading that imaginary line from right to left. The Sedins, Burrows, Kesler, Samuelsson, and so on for forwards, and Hamhuis, Bieksa, Edler, and so on for defencemen.

A player's position relative to that imaginary line can also tell us something about how trusted he was defensively. If a player is above that imaginary line, then he got more ice time than could be justified based on his scoring. It could mean that the player was strong defensively. In this case, that assessment certainly applies to Kesler and his winger on the top shutdown line, Mikael Samuelsson.

It's unfortunate that the top defencemen are all bunched together like that, though. Based on the original player usage chart, and even on shots per game, we know that Ehrhoff (2.4) was used more offensive mindedly than Hamhuis (1.6), but that distinction is lost in these new charts.

Overall, I'd rate this a B–. We can get an accurate picture of a team's depth chart and a sense of which players performed better than others, and we can identify a few players whose usage stood out, but it doesn't point out the defensive players quite as clearly, nor help us sort out the blue line.

With that quick, understandable example under our belts, let's go back in time and consider a prominent team for which there is no modern player usage chart.

The 2000–01 Colorado Avalanche

Individual ice-time statistics were first recorded in the 1997–98 season, which means that the 2000–01 Colorado Avalanche was one of the first Stanley Cup champions that can be studied using at least three years' worth of ice-time data.

The 2000–01 Avalanche has the added advantage of being a particularly memorable team, thanks to the presence of several future Hall of Famers, like Joe Sakic, Peter Forsberg, Ray Bourque, and Rob

Blake. We should be able to partly evaluate the following simplified player usage chart even without a regular version for comparison.

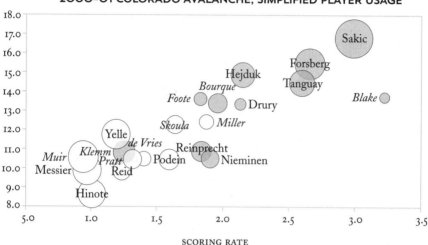

2000–01 COLORADO AVALANCHE, SIMPLIFIED PLAYER USAGE

Well, Sakic, Forsberg, and Bourque are exactly where they are expected to be, which should help place all of the other players in context.

Blake is in an unusual spot, however. He is among the team's leaders in ice time, but not to the extent that his scoring would suggest. All things being equal, a player with his enormous offensive contribution should be getting a lot more ice time and, therefore, located much higher on the chart than the team's other defencemen. But all things are not equal. Blake was acquired at the NHL trade deadline, so his scoring rate is based on how he performed in Los Angeles, not in Colorado. Plus, the two players used as often as Blake were Adam Foote, who was an acknowledged defensive specialist, and Ray Bourque, who is one of the most accomplished defencemen in history. It makes perfect sense that Blake didn't get significantly more ice time than they did, no matter how much he scored.

To refresh my memory on the balance of the roster, I spoke with author and noted Avalanche expert Adrian Dater. He recalled that the Avs basically rolled their lines, without preference to one line or another based on the score or game situation. Sakic, Alex

Tanguay, and Milan Hejduk formed the top scoring line, Forsberg was amazing on an often-changing second line with Chris Drury and others, while Stéphane Yelle, Shjon Podein, and shot-blocking master Eric Messier formed the defensive-minded third line.

To interpret the simplified player usage chart, we can apply what we learned by studying Vancouver and visualize a line that goes from the bottom left of the chart to the top right. Going from right to left, it appears to accurately read like the team's depth chart.

As explored with Vancouver, anyone located above that imaginary line was getting more ice time than their scoring warranted. Absent another explanation, it's possible that the extra ice time was a result of stronger defensive play. It seems to confirm Dater's view on Yelle and on how the four lines were rolled. Hejduk is also vertically higher than expected, which seems to be an accurate perspective of his play, given that he received some Selke consideration that season.

Similarly, those located slightly below that imaginary line got less ice time than their scoring warranted, and that's potentially because they were used a bit more carefully and not as trusted in tough defensive situations. From this perspective, it makes sense that such players include rookies Steve Reinprecht and the pesty Ville Nieminen. The fact that they have nicely shaded circles could be a consequence of being somewhat sheltered and/or of occasionally playing on the second line with Forsberg.

The other interesting shaded circle belongs to Jon Klemm, who is mixed in with Colorado's other depth defencemen. It appears he would have been effective in a more challenging role on a team without three legends ahead of him on the depth chart. And, sure enough, he went on to play an effective top-four role as Chicago's defensive specialist the following two seasons.

While this wasn't exactly a deep dive, this simplified player usage chart certainly seems to have quickly painted a reasonably accurate and illuminating picture of the 2000–01 Avalanche and how their players were used. In turn, it could be used to put some of the team's stats in context. I'll give that a solid B.

Now let's see if these charts are useful for placing data in context when we step outside of our comfort zone.

For all but the NHL's newest fans, knowledge about the 2010–11 Vancouver Canucks and 2000–01 Colorado Avalanche is quite strong, but what if we considered a team that we didn't know at all? Can we use these simplified player usage charts to understand how unfamiliar teams deploy their players and to put their results in context?

Other than possibly Finland's SM-liiga, the Swedish hockey league is the only league that has all of the required statistics available to build these simplified charts. Based on the experience gained from reading the previous two charts, consider the following chart for the 2014–15 Frolunda Hockey Club and try to determine how each unfamiliar player was used.

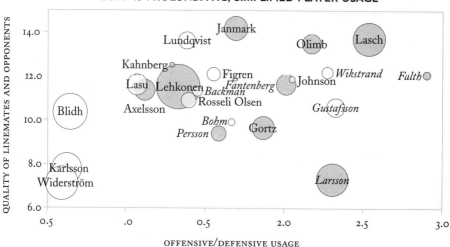

2014-15 FROLUNDA HC, SIMPLIFIED PLAYER USAGE

Given their position above the imaginary diagonal line, Mattias Janmark and Joel Lundqvist appear to be the top defensive players, with Mathis Olimb and Magnus Kahnberg close behind. Max Gortz is really the only forward below that line, meaning that he was probably used for secondary scoring and was not completely trusted with tough defensive situations.

The remaining forwards appear to be used in standard fashion, either on a top line, like Ryan Lasch and possibly Andreas Johnson, or in a more secondary role, like Robin Figren, Mats Rosseli Olsen, Anton Axelsson, and Nicklas Lasu. Artturi Lehkonen is one of the intriguing names on the chart, given that he's used as a typical secondary forward like those latter four, but with tremendous plus/minus-based results. To me, that looks like a top prospect who is being developed quite carefully.

Among the defencemen, we know that it will be hard to make distinctions among the top four, which appear to be Elias Falth, Mikael Wikstrand, Oscar Fantenberg, Christian Backman, and possibly Erik Gustafsson. Falth is likely the team's Blake-type forward, Backman is the defensive specialist, and Fantenberg probably gets to play with the top line quite often. Oliver Bohm and Christoffer Persson are clearly third pairing options.

The most interesting player on the blue line is Jacob Larsson, who appears to be used rather carefully but with fantastic results. Absent any other information, he appears to be a carefully developed prospect as well.

But, of course, we do have other information, since two full seasons have passed, and we have learned a lot about each player since then:

- Janmark went on to play for the Dallas Stars the following season, where he capably played in a somewhat offensive-oriented third-line role.
- Gustafsson played as a seventh defenceman for Chicago the following season, and he has otherwise served as perhaps the top two-way option for their AHL affiliate, the Rockford Ice Hogs.
- Gortz was the second-leading scorer for Nashville's AHL affiliate, the Milwaukee Admirals, before struggling the following season and being dealt to Anaheim's affiliate, the San Diego Gulls.
- Anton Blidh played for Boston's AHL affiliate, the Providence Bruins, before spending some time on Boston's depth lines.
- Erik Karlsson played on the depth lines for Carolina's AHL affiliate, the Charlotte Checkers (and he has not been a star defenceman for the Ottawa Senators).

- Johnson, who had been drafted in the seventh round by the Toronto Maple Leafs in 2013, played for their AHL affiliate, the Toronto Marlies, two years later.
- Larsson was indeed a highly touted prospect and was drafted by Anaheim 27th overall in the 2015 NHL Entry Draft.
- Finally, Lehkonen is also another great prospect, had been drafted by Montreal 55th overall in the 2013 draft, and has since moved on to the Canadiens, where he has spent some time on the top two-way line with Tomas Plekanec.

Also, we know that Lundqvist and Backman had previously played in the NHL, and it makes a lot of sense that they would be among Frolunda's most trusted defensive players in their later years.

At the time, I checked the chart and my interpretation with Petter Carnbro, who worked for Frolunda HC, and he felt that "overall, I think your assessments are quite accurate." One exception is Wikstrand, whom Carnbro felt was already an NHL-ready top-pairing defenceman. In fact, he would have played for the Ottawa Senators by now, if not for some personal issues that kept him at home.

Carnbro's only other concern was that there are clear differences among the secondary forwards that couldn't be captured in this chart. Specifically, he identified Figren and Kahnberg as point producers and Rosseli Olsen, Lasu, and Axelsson as defence oriented.

So while the chart wasn't perfect, it clearly identified the correct role for most of the team's players, and it did correctly identify the team's more outstanding players. And yes, that does appear to be the kind of information that would help put player data in the right context when translating it to an NHL standard. Let's call this one a B+.

The 1976–77 Montreal Canadiens

As one final experiment, let's look at arguably the greatest NHL team of all time, the 1976–77 Montreal Canadiens. Their lineup included eight future Hall of Famers—11 if you include goalie Ken Dryden, coach Scotty Bowman, and GM Sam Pollock. Four of the six members of the First All-Star Team that year were Canadiens, they swept virtually every individual and team award, and when you consider the regular season and playoffs together, they won 72 games, lost 10, and tied 12 while outscoring their opponents 441-193. It was a truly amazing team.

Unfortunately, we don't have any ice-time data for the 1976–77 Habs. Just like the non-Nordic European leagues, the AHL, all U.S. college leagues, and the Canadian major junior hockey leagues, the NHL's history doesn't have the data required to make even the simplified player usage charts.

There are ways to estimate ice time, as detailed in the *Hockey Compendium* back in 2001.[49] It involves using the goals that were scored (for and against) while a player was on the ice, which was first published in the 1999 book *Total Hockey*.[50] My friend and frequent co-author, avid hockey historian Iain Fyffe crunched the numbers and provided the estimates used in the following chart.

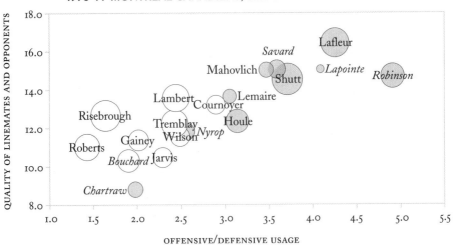

1976–77 MONTREAL CANADIENS, SIMPLIFIED PLAYER USAGE

Reading this chart from right to left along the imaginary diagonal line does result in a reasonably accurate depth chart. Seven of

49. Jeff Z. Klein and Karl-Eric Reif, "Players," *The Hockey Compendium: NHL Facts, Stats, and Stories* (Toronto: McClelland & Stewart, 2001) 51–90.

50. Dan Diamond, "Modern Player Register," *Total Hockey* (New York: Total Sports, 1999), 824–1600.

the eight players to be later named to the Hall of Fame are among the first eight players.

The one player at the top of the line up who is not in the Hall of Fame is 6-foot-5 giant Pete Mahovlich. This chart is based on three seasons' worth of data, but 1976–77 was the season that he was transitioned off the dominant top line with Guy Lafleur and Steve Shutt in favour of Jacques Lemaire. Instead, Mahovlich played on more of a two-way line with Yvon Lambert and team captain Yvan Cournoyer. That certainly places his scoring in context, which dropped from 117 points in 1974–75 and 105 points in 1975–76 to 62 points in 1976–77.

The one future Hall of Famer who isn't at the top of the depth chart is Bob Gainey. Given that he won the Selke Trophy as the league's best defensive forward for four consecutive seasons after it was introduced in 1977–78, it's a common belief that he was on the top defensive line in 1976–77, too. This may look like an error in the chart, but this chart encompasses Gainey's age 21 through age 23 seasons, in which he generally played on more of a secondary checking line with Doug Jarvis and Réjean Houle. Even if you don't trust our ice-time estimates, the fact that his shots per game increased from 1.8 in 1976–77 to 2.1 in 1977–78 and that his penalty minutes per game increased from 0.51 to 0.86 suggest that he was still working his way up the depth chart at this time.

As for the blue line, teams generally played with four defencemen in those days, although Montreal appeared to be more of a three-player operation. Larry Robinson was known to play on the top pair with Serge Savard, leaving offensive specialist Guy Lapointe to work with Pierre Bouchard and Bill Nyrop.

By and large, it isn't easy to evaluate Montreal's player usage using ice-time estimates, since they have an impact on the horizontal axis and each player's even-strength scoring rates are also based on those estimates. In general, that's why even these simplified player usage charts won't be of tremendous insight for any league that doesn't record individual players' ice time.

In my view, that's about as far as we can go in our examination of how player usage charts can be simplified to look at other leagues and eras and place their stats in context. Let's hop in our DeLorean, travel back to the future, and consider how new developments can refine the accuracy and usefulness of player usage charts in the modern-day NHL.

New Developments in Visualizations

For most fans, their first experience in the world of hockey analytics is a table full of figures and strange acronyms that are impossible to place in context. One player is a 35.4 in one metric and a −7.0 in another, while another player is a 55.8 and a 12.0. What do those numbers mean? What's good and what's bad? Even for experienced number-crunchers, it isn't easy to quickly process a table full of figures and place them in context.

That's one reason why player usage charts were embraced so quickly. To some, offensive zone starts, quality of competition, and relative Corsi are just confusing numbers when presented separately and/or in a table of figures, but they instantly click when presented together in a chart.

That's why player usage charts' most lasting legacy might be how they marked the transition from a world almost exclusively filled with tables of figures to one that is increasingly populated by visual representations of the data. Now, the first experience for many fans is some fancy graphics. That's a far easier way to get hooked and to start putting data in context.

At first, the only hockey visualizations were the NHL's shift charts, introduced in 2002–03, and heat maps of all of a player's or team's shots and/or goals.[51] I'm not sure who made the first heat maps, but my first experience included those on Sporting Charts (sportingcharts.com/nhl) and my friend Greg Sinclair's Some Kind of Ninja website, which is the now-defunct predecessor to his Hockey Stats website (hockeystats.ca).[52] Since then, these heat maps

---

51. An example of an original NHL shift chart from 2002–03 can be found at http://www.nhl.com /scores/htmlreports/20022003/SCH21230.gif.

52. The Some Kind of Ninja website was www.somekindofninja.com, but it has been off-line since 2015.

─ COACHING STAFF ─

HEAD
COACH

POWER PLAY
COACH

FIGHTING
COACH

have been made available at a number of different sites, most notably at Natural Stat Trick (naturalstattrick.com).

Greg's site also hosted the first interactive player usage charts tool, which predates the version added to my own Hockey Abstract website by the venerable visualization expert Robb Tufts.[53] Eventually, player usage charts tools could be found on most major websites, like Progressive Hockey (progressivehockey.com) by Matt Pfeffer, who is now with the Nashville Predators; Extra Skater by Darryl Metcalf, who is now with the Toronto Maple Leafs; its spiritual successor War on Ice by Andrew Thomas and Sam Ventura, now of the Minnesota Wild and the Pittsburgh Penguins; and that site's spiritual successor, Emmanuel Perry's Corsica Hockey.[54] (Apparently, building these tools is a pathway to a front-office position.)

Since then, the community has built on these three initial types of visualizations and have gone in entirely new and creative directions. It started with Ben Wendorf, who launched the now-inactive Hockey Visualized website, which combined his love of hockey history with

---

53. An example of Greg's player usage chart tool can be found at www.fearthefin.com/2014/8/20/6046829/with-extra-skater-dead-heres-where-you-can-get-your-advanced-stats-fix; Robb Tufts's player usage chart tool is available on my website, at www.hockeyabstract.com/playerusagecharts.

54. The Extra Skater website was www.extraskater.com, but it has been off-line since 2015. An example of its player usage chart tool can be found at https://nesn.com/2014/04/2014-nhl-awards-tracker-patrice-bergeron-leading-selke-trophy-race/. The War on Ice website was www.war-on-ice.com, but it has been off-line since 2016. An example of its player usage chart tool can be found at http://nerdhockey.com/by-the-numbers/2015/10/2/tanner-glass-still-in-the-nhl. The Corsica Hockey website was www.corsica.hockey, but it has been off-line since 2017. An example of its player usage chart tool can be found at www.fearthefin.com/2017/5/4/15540504/2016–17-season-in-review-patrick-marleau.

his natural talent for visualizations, and who co-launched Hockey Graphs (hockey-graphs.com), which has become one of the leading sources of statistical hockey analysis.[55]

Then there's Domenic Galamini, whose HERO (formerly WARRIOR) charts have become an indispensable way of placing some of today's leading stats in context by comparing them graphically to another player, to the player's position in a depth chart, or to a statistical archetype.[56]

At last, our history lesson ends with the incredible Micah Blake McCurdy, who has taken over as our field's new ambassador for hockey visualizations (and subsequent analysis). At the time of writing, McCurdy's HockeyViz website (hockeyviz.com) is home to a few dozen original visualizations, including divisional standings, playoff chances, draft lottery probabilities, line combinations, the scoring race, team statistics, on-the-fly shift changes, scoring networks, skater context charts (which are his own take on player usage charts), and virtually anything else that such a brilliant mind can possibly conceive. It is the go-to destination for hockey visualizations, and it is financially supported voluntarily by the community through Patreon to the tune of over $2,000 per month, as of September, 2017.[57] That may not be quite enough to quit your day job, but it easily places him among the most financially successful members of our community.

Specific to this chapter's quest to improve player usage charts, McCurdy unveiled score-state deployment charts halfway into the 2015–16 season as a graphical portrait of which players teams use based on the game-score situation. Specifically, they depict what percentage of ice time is assigned to each forward or defenceman when the team is tied, when leading by one, two, three, or more goals, or when trailing by the same. It is an excellent way to determine which

---

55. The Hockey Visualized website was www.hockeyvisualized.com, but it has been off-line since 2017.

56. Domenic Galamini, "HERO Charts – Frequently Asked Questions," *Own the Puck* (blog), January 21, 2017, https://ownthepuck.wordpress.com/2017/01/21/hero-charts-frequently-asked-questions/.

57. I encourage you to become a patron of HockeyViz through Patreon at www.patreon.com /hockeyviz?ty=h.

players are used to protect tight leads and which players are used to chase them.

At my request, McCurdy was kind enough to put together some specific scor- state deployment charts for the 2010–11 Vancouver Canucks. Let's start with the defencemen and compare them to the player usage chart from earlier in this chapter.

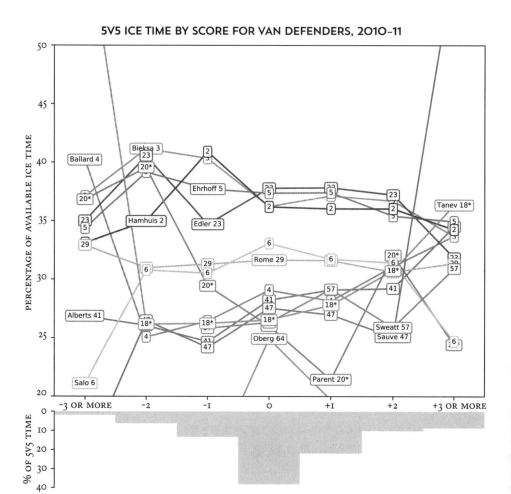

5V5 ICE TIME BY SCORE FOR VAN DEFENDERS, 2010–11

© Micah Blake McCurdy, IneffectiveMath, hockeyviz.com

Let's use Hamhuis as an example of how to read this chart. When the Canucks were down by two goals (−2), Hamhuis got about 35% of the available ice time, which places him fourth on the depth chart. Following his line across the chart (jersey No. 2), it rose to over 40% and the team's top choice when down by one goal, and then back down to around 37% when the score was tied (0) or when the Canucks were up by one (+1) or two (+2) goals, which generally places him third or fourth in a tight pack.

So what does this mean? When Vancouver was trailing the game by a single goal, they would play the defencemen whose offensive abilities they trusted the most, independent of their defensive abilities. In 2010–11, Hamhuis and Kevin Bieksa got over 40% of the ice time in such situations, followed by Christian Ehrhoff at around 38% and Alexander Edler at 35%. That suggests that, for the offense, coach Vigneault's trust was greatest in Hamhuis.

Within the top four, that scenario is actually the opposite of what the player usage chart suggests. If you recall (or flip back a few pages), Ehrhoff and Edler were used more frequently in the offensive zone and against lesser competition, while Hamhuis and Bieksa were taking on top opponents in more of a two-way role. I spoke with my friend Jim Jamieson, who covered the Canucks for the *Province* that season, and he agrees that Ehrhoff and Edler were the offensive-minded pair, not Hamhuis and Bieksa.

Defensively, the team's most trusted players would be deployed when the team was protecting a one-goal lead (+1). From this perspective, there doesn't appear to be any separation between those top-four defencemen. Despite their position on the player usage chart, they all appeared to be equally trusted to protect a tight lead.

As for the team's other options, Sami Salo and Aaron Rome were clearly third-pairing choices who got about 32% of the ice time in any given game situation. Christopher Tanev, Andrew Alberts, and Kevin Ballard fit the profile of depth players, being assigned less than 30% of available ice time, except when the team was blowing away its opponents by three goals or more.

The original player usage chart makes Salo and Ballard appear to be more prominent players, but that's partly a function of the fact that it uses three seasons' worth of data, instead of the single year used in the score-state deployment chart. In essence, Salo and

Ballard were once top-four players but were only used that way in 2010–11, when one of the regular top-four defencemen was injured (he missed a combined 68 games to injury).

The score-state deployment chart for forwards is a little busier, but it's clear enough to interpret in roughly the same way as the defencemen.

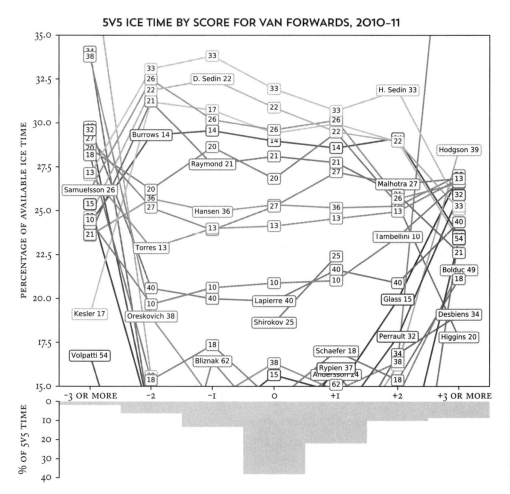

© Micah Blake McCurdy, IneffectiveMath, hockeyviz.com

From this perspective, the forwards used most frequently when the team needed a goal (−1) included Daniel and Henrik Sedin (obviously) followed by Ryan Kesler, Mikael Samuelsson, Alexandre

Burrows, Chris Higgins, and Mason Raymond. When protecting a late lead (+1), the Sedins were not really used more often than any of the other five forwards, and Manny Malhotra was thrown into the mix. Unlike our analysis of defencemen, none of this information is really at odds with the player usage chart.

The information in these score-state deployment charts offers a different perspective on the offensive-vs-defensive deployment of a team's players. When they agree with zone start percentages, then we can be reasonably confident about how a player was used and, therefore, how to place his data in context. When they differ, we have identified players whose usage may warrant a closer look.

This score-state information can be easily integrated into a player usage chart. Since the horizontal axis is meant to represent whether a player is used in more of an offensive or defensive capacity, the extent to which players are used to chase or defend leads is a perfect fit.

Currently, a player's offensive deployment is measured using zone start percentages, which is the share of shifts a player starts in the offensive zone rather than the defensive zone. Since it ignores on-the-fly changes and faceoffs that occur in the neutral zone, it is setting aside a lot of data.

In contrast, score-state deployment charts can use every minute of data, which is always a big step toward achieving more accurate stats. It's also possible that the rate at which players are called upon to chase and/or protect a lead is a better indication of which players are used offensively than the zones in which they start their shifts.

Also, consider how much more clearly the score-state deployment charts describe the usage of secondary players. While Kesler is at the top of the chart when protecting a one-goal lead, those who actually appear to his left on the player usage chart, like Jannik Hansen, Manny Malhotra, and Tanner Glass, are correctly shown as secondary options on the score-state deployment chart. In essence, some players appear to be defensive minded not because they are trusted defensively, but because they aren't used offensively at all.

There are a few ways to integrate this information into player usage charts. The simplest way is to take the difference between a player's share of the ice time when the team is trailing and when it is leading. In fact, that's exactly the method used by Alex Novet, who independently started exploring this exact same idea before I had made any of these ideas public.[58]

However, my approach is to express this information as a percentage. After all, there is a difference between someone who gets 35% of the ice time when trailing and 29% of the ice time when the team is leading, and someone who gets 15% and 9%, respectively. Even though they would both score as +6% in absolute terms, the latter's ice time is far more skewed to offensive purposes. In the following chart, the former fictional player's offensive deployment percentage would be 35 divided by the sum of 35 and 29, which works out to 54.7%, while the latter's would be 62.5%.

Using this data in place of zone start percentages but keeping everything else the same, here's how the revised 2010–11 Vancouver Canucks player usage chart looks.

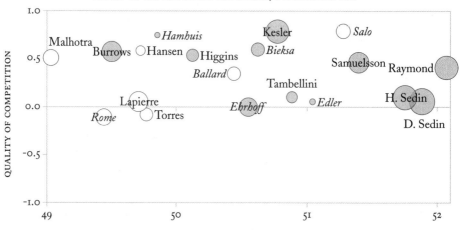

2010-11 VANCOUVER CANUCKS, PLAYER USAGE

PERCENTAGE OF ICE TIME WHEN TEAM IS TRAILING

Is this a better perspective? To summarize the differences, Burrows has shifted significantly to the left, followed by defencemen Rome and Ehrhoff, with slight moves by Lapierre and Kesler. Offensively, Raymond made the only significant shift, followed by lesser moves by

---

58. Alex Novet, "An Alternative to Player Usage Charts," *Fancy Stats Island* blog), January 14, 2016, https://fancystatsisland.wordpress.com/2016/01/14/an-alternative-to-player-usage-charts/.

Samuelsson and Salo. There was not enough data to include Alberts or Glass on the chart. Perhaps it's a matter of personal opinion, but that seems like a slightly more accurate representation to me.

What did Jamieson think of these differences? He definitely confirmed that Burrows was a defensive-minded player, who may only have been on the top line to serve as a physical presence for the Sedins. Likewise, he absolutely confirmed that Raymond was an offensive-minded talent who was often placed on the shutdown line with Kesler and Samuelsson to give them an offensive spark and because they could help compensate for any of his defensive errors. Jamieson was quite pleased with this chart from that perspective.

However, he preferred Ehrhoff's position on the original chart. While he was pleasantly surprised by his responsible defensive play, Ehrhoff was clearly an offensive-minded defenceman who should be on the right side of the chart, with the Sedins, and not in the middle. Even when using traditional stats, Ehrhoff averaged 185 shots and 45.3 points per season over this time period.

Ehrhoff is also a good cautionary tale about the real-world dangers of failing to place a player's stats in context. That summer, the Sabres signed Ehrhoff to a front-loaded 10-year, $40 million contract. After he struggled to perform in a more complete two-way role, the Sabres bought out that control three years later, despite owing him just $18 million for the final seven seasons.

With this one exception, Jamieson was pleased with the accuracy of the slight differences with this new approach to the horizontal axis. As for the vertical axis, it can be replaced whenever a more accurate estimate of the average quality of a player's opponents and/or linemates is developed. How to best measure quality of competition and/or linemates and the importance of the results are still open problems. At the time of writing, new methods have been or are being actively developed by WoodGuy, Emmanuel Perry, and Tyler Dellow.[59]

---

59. WoodGuy, "WoodMoney: A New Way to Figure Out Quality of Competition in Order to Analyze NHL Data," *Woodblog*, July 18, 2016, http://becauseoilers.blogspot.ca/2016/07/woodmoney-new-quality-of-competition.html; Emmanuel Perry, "Bootstrapping QoT/QoC and the Sedin Paradox," *Corsica Hockey* (blog), August 22, 2016, http://www.corsica.hockey/blog/2016/08/22/bootstrapping-qotqoc-and-the-sedin-paradox/; Tyler Dellow (@dellowhockey), Twitter, March 22, 2017, 9:52 a.m., https://twitter.com/dellowhockey/status/844592594135011329.

In addition, the shaded circles could be replaced with score-adjusted variations of shot-based metrics, which account for the different ways that teams play when protecting or chasing a late lead and how that affects their numbers.[60] Another possibility is to use expected goals, which places a different weight on each shot attempt based on its chances of going in.[61] That calculation is based on the location of the shot attempt, among other quality factors.

There's no limit to the possible ways of improving player usage charts over the long run. The individual metrics may change over time, but the purpose of player usage charts is constant, which is to graphically present how a team's players are being deployed, which helps put their data in the proper context.

Closing Thoughts

Context is everything in the world of statistics, and player usage charts help put the numbers in context. However, despite the importance of context, remember that no statistical perspective is perfect. Even when player usage charts are simplified for use in other leagues and eras, or when they're updated with the latest developments, they're still not a flawless representation of how players are used—and they never will be.

Always keep a player's assignment in mind when making judgments about his performance, either with statistics or through traditional analysis. In that regard, player usage charts are a useful tool to help put a player's performance in the proper perspective and to gain insight into how that player's performance has been impacted by the assigned role on the team.

60. Eric Tulsky, "Score-Adjusted Fenwick," *Broad Street Hockey* (blog), January 23, 2012, https://www.broadstreethockey.com/2012/1/23/2722089/score-adjusted-fenwick.

61. Alan Ryder, "Shot Quality," Hockey Analytics, January 2004, http://hockeyanalytics.com/2004/01/isolating-shot-quality/.

WHO IS THE BEST WOMEN'S
HOCKEY PLAYER?

Watching the coverage of Hayley Wickenheiser's retirement announcement in January 2017 was highly inspirational. Given the tremendous growth in popularity of women's hockey since Wickenheiser first donned a jersey, are there many active players who have had a greater impact on our beloved sport?

Women's hockey has come a long way since Wickenheiser laced up her first pair of skates. According to Hockey Canada, a 12-year-old Wickenheiser was one of only 8,146 women playing hockey in the country in 1990–91.[62] And according to the IIHF 2016 Survey of Players, there were 87,500 women playing hockey when she retired.[63] That's over a 10-fold increase!

I'm almost exactly three years older than Wickenheiser, and I can only recall a single girl who played hockey when I was a child. (Hi Kerry!) For every girl who played hockey when I was a boy, there's almost an entire team of girls playing today. That's probably why, as an adult, I can name dozens of women who have shared the ice with me in various recreational leagues around Calgary and why I see several girls per team whenever I watch my young nephew play.

---

62. Data from "Statistics & History," Hockey Canada, updated May 16, 2017, http://www.hockeycanada.ca/en-ca/Hockey-Programs/Female/Statistics-History.aspx.

63. Data from "Survey of Players," International Ice Hockey Federation, updated May 16, 2017, http://www.iihf.com/iihf-home/the-iihf/survey-of-players/.

Of course, Wickenheiser isn't alone in having inspired a generation of Canadian girls to play hockey. These same inspirational sentiments were expressed and shared upon the retirements (or semi-retirements) of stars like Caroline Ouellette, Jayna Hefford, and Jennifer Botterill. And the popularity of women's hockey has had a similar explosion in the United States and is well underway in the Nordic countries, in Russia, and elsewhere.

Even though the growth of this sport would be an interesting topic in its own right, this chapter is about identifying who is the world's best women's hockey player. In her prime, there was little question that Wickenheiser was number one, but who is it today?

To answer this question, I started by looking at the various league websites, to see what kind of data there was to work with. Things looked pretty grim. As I began to see the parallels between the data available for women's hockey today and men's hockey in the early years of its development, it became clear just how difficult it would be to answer this seemingly simple question.

As with men's hockey a century ago, there's a very small pool of data with which to compare players, and it is spread across many different leagues and tournaments, each of which involves very few games, records a very limited selection of statistics of unreliable accuracy, and is subject to the skewing effects of the highly variable quality of teams, leagues, and individual players.

In women's hockey, the closest equivalents to the NHL in terms of being a top professional league are the National Women's Hockey League (NWHL) and the Canadian Women's Hockey League (CWHL), which have a combined total of nine teams and play just 18 and 24 games per season, respectively.[64] That's not a lot of data.

Furthermore, the lack of decent pay (or *any* pay, in the case of the CWHL prior to 2017[65]), means that the world's best player may simply not be in a position to play in those leagues. It's also pos-

64. At the time of writing, they are expanding to a combined total of 11 teams. The CWHL is adding two teams in China, the Kunlun Red Star and the Vanke Rays.

65. Rachel Brady, "Canadian Women's Hockey League Will Begin Paying Its Players," *Globe and Mail*, September 1, 2017, https://beta.theglobeandmail.com/sports/hockey/cwhl-will-pay-its-players-for-the-first-time-starting-this-season/article36139819.

sible that today's Wickenheiser is still in school, playing U Sports (formerly Canadian Interuniversity Sport, or CIS) hockey as Wickenheiser once did or NCAA Division I hockey in the United States. Setting aside national pride, it's also possible that the world's best player is actually somewhere in Europe.

Regardless of where she normally plies her trade, the world's best women's hockey player will almost certainly be a part of the annual World Championship or the most recent Winter Olympics. However, that just adds five games of data to the mix per year, and some data will get skewed by the wildly variable quality of opposing teams. We could supplement with data from other leagues, but the relative lack of both players and games makes the translation process far less reliable than the method laid out for the NHL.

Even if the various scraps of data are gathered correctly, the greater problem is the inability to put those stats in context, whether through player usage charts or otherwise. Stats beyond simple goals and assists are almost impossible to come by, and even those traditional stats can be hard to track down for certain leagues and seasons.

Quite frankly, trying to statistically establish the world's best women's hockey player is probably futile with the available information. But for every journey, there is a first step. We will simply have to take the strongest stride possible today and hope that we can follow up with some stronger ones in the future, much like Wickenheiser herself did. Let's begin!

The Subjective View

As always, after a question has been defined, the next step is to establish the subjective answer, to use as a sanity check when the work is complete. (Or a "sniff test," as many of my colleagues prefer to more colourfully describe it.) Doing this up front prevents that sanity check from being influenced by the statistical perspective that follows.

Since Wickenheiser's prime, there hasn't been a consensus on the world's best women's hockey player. Recently, the debate has centred around Marie-Philip Poulin of the CWHL's Montreal Canadiennes and either Hilary Knight or Brianna Decker of the NWHL, who are long-time teammates of the Wisconsin Badgers, Boston Blades, Boston Pride, and U.S. national team. When I first started this chapter in February 2017, a quick Twitter poll had Poulin in front with 50% followed by Knight with 33%.[66] Decker was at 7%, and 10% had someone else in mind (including me). When I spoke with some of the women's hockey players whom I know, their most common answer was also "Pou" (which, spelled differently, is quite similar to my own hockey nickname).

As a Canadian, I've had plenty of opportunity to watch Poulin play and completely understand her incredible popularity. Like Wickenheiser, she was a star at a very early age. In 2007–08, she was runner-up for the CWHL Player of the Year while playing with the Montreal Stars at age 16, and she also made her first appearance on the Canadian national team, where she later scored the gold-medal-game-winning goal at both the 2010 and 2014 Winter Olympics.

Even American fans got a look at her incredible talent in her four seasons with Boston University, which culminated in finishing runner-up for the Patty Kazmaier Award, as the league's top player, in 2014–15. She then played for the Montreal Canadiennes in 2015–16 and won practically every award and distinction the CWHL had to offer, including the Jayna Hefford Memorial Trophy for the player-voted MVP and the Angela James Bowl for the league scoring leader.

Knight also stands out, both for her skill and her size—at 5-foot-11 and 172 pounds, she is one of the more physically imposing players in the women's game today. She scored 262 points in 161 games for the Wisconsin Badgers, 62 points in 41 games for the CWHL's Boston Blades, led the NWHL in scoring with 33 points in 17 games for the Boston Pride in 2015–16, and has 71 points in 50 combined games in World Championships and Olympics.

Spending most of her career with Decker might have resulted in a bit of a mutual boosting effect that will be difficult to straighten out. Playing alongside Knight, Decker scored 246 points in 143 games for the Badgers, winning the Kazmaier Award in 2012. Jared

---

66. Rob Vollman (@robvollmanNHL), "Who is the world's best women's hockey player?," Twitter, February 3, 2017, 12:31 p.m., https://twitter.com/robvollmanNHL/status/827615490843373568.

Clinton of *The Hockey News* named her as the world's best women's hockey player in September 2015.[67]

My own subjective preference is Meghan Agosta. In my local charity hockey work, I have had the opportunity to share the ice with Agosta, whose jaw-dropping skill is matched only by her generosity. In college, Agosta scored an incomprehensible 302 points in 134 games for the Mercyhurst Lakers, finishing as a finalist for the Kazmaier all four of her seasons there. She then set the single-season CWHL record with 80 points in 27 games for the Montreal Stars in 2011–12, and she led the league again with 46 points in 23 games the following season. Representing Canada in international competition since 2006, Agosta has won three Olympic gold medals and was the MVP for the 2010 games. Unfortunately, the financial reality of women's hockey means that her training and subsequent day job with the Vancouver police has prevented her from playing regularly, which may start affecting her play now that she's in her 30s. I'll concede that she may no longer be part of the discussion.

To be fair, there are many others who could be mentioned, but as long as our results include Poulin, Knight, and Decker, then they probably have some merit. Let's begin.

The World Championship

It stands to reason that the world's best women's hockey player, whomever she may be and no matter where she lives, has competed in the Olympics. However, there are some particularly notable problems with using Olympic data to evaluate players:

- Only basic scoring data is available, like goals and assists.
- The wildly varying quality of the teams can skew a player's results based on both her teammates and which teams she faced.
- There are only five or six games' worth of data.
- Since the Olympics are only once every four years, the world's

---

67. Jared Clinton, "Top 10 Women's Players from Last Season," *The Hockey News*, September 27, 2015, http://www.thehockeynews.com/news/article/top-10-womens-players-from-last-season.

best player could be different from one tournament to the next, and it could possibly be someone else entirely in between them.

These are not unsurmountable problems. We'll start by tackling the last two concerns, which we can partially address by using all data from every annual IIHF World Championship instead of just the Olympics.

Theoretically, both World Championship and Olympic data should be of the same quality, given the similarity in rules, tournament format, and the players and teams involved. Statistically, the correlation between individual players' data from the Olympics and World Championship is 0.82, based on the 16 players who have competed in at least 20 games in both since 1998. The difference in their average scoring rate is pretty small, 0.77 points per game versus 0.81.

If we add up everyone's scoring totals from 1998 to the present, the accumulation of data for those who have played the longest starts to resemble an 82-game NHL season, which helps us to start putting these numbers in context. For example, there are two players who have topped 50 goals and 100 points in roughly 80 career games, Wickenheiser and Hefford. Those are achievements that are easy to understand.

TOP SCORERS IN WORLD CHAMPIONSHIPS, 1998 TO 2017[68]

PLAYER	TEAM	GP	G	A	PTS
Hayley Wickenheiser	Canada	76	51	76	127
Jayna Hefford	Canada	81	52	57	109
Caroline Ouellette	Canada	79	32	62	94
Jenny Potter	USA	71	34	59	93
Natalie Darwitz	USA	55	43	40	83
Hilary Knight	USA	55	40	37	77
Jennifer Botterill	Canada	61	30	44	74
Krissy Wendell-Pohl	USA	39	25	44	69
Danielle Goyette	Canada	46	38	30	68
Brianna Decker	USA	35	26	38	64

---

68. Data for World Championship from Elite Prospects, accessed May 16, 2017, http://www .eliteprospects.com.

The fact that Wickenheiser ranks first certainly lends some credibility to the idea of using this data to evaluate hockey players, but this leaderboard also confirms our initial concerns.

First, there is still not a lot of data. For Wickenheiser, 76 games seems like a lot, but since it's spread out over a long time frame, we can only establish that she *was* the best player and cannot offer any evidence of who may be the best player today. However, if we tried to address this by looking at a weighted average of each player's past three seasons, then we'd have only 15 games per player. Anybody can get hot for 15 games.

Furthermore, Decker and Wendell-Pohl scored more points on a per-game basis, 1.83 and 1.77, than Wickenheiser did, 1.61. Does that mean they were more effective? At any level of hockey, it can be challenging to extrapolate what one player's performance over 35 or 39 games would look like over 76 games. You can't just assume that the existing scoring rate will continue, and it's not always easy to calculate at what rate it might drop.

The easiest way to deal with the lack of data is to study each player's performance in other leagues as well, such as the various professional leagues and the college and university leagues. Once adjusted for league quality, we can add in many more games' worth of data, which makes it possible to focus only on the past three seasons.

Before we can proceed with that strategy, we need to devise a process to address the concern that all of the players on that leaderboard come from one of two talent-laden teams, Canada and the USA. What if the world's best hockey player is someone like Finland's Michelle Karvinen? She has scored 50 points in 50 games in World Championships, which may pale by comparison to the athletes in the earlier table, but she didn't exactly have linemates like Wickenheiser or Wendell-Pohl padding her stats. What would Wickenheiser's numbers have looked like if she had been born in Helsinki instead of Shaunavon, Saskatchewan?

As explored in the opening chapter, the team for which a player competes is one of the key factors that can influence a player's scoring. We need to account for the greater scoring opportunities that certain players have in order to compare players across teams.

In order to address this, we can judge players based on their scoring relative to team-based expectations, rather than on the base results. That is, we can first calculate how many points a player on the same team and of the same position would be expected to score, based on the combined sum of a player's teammates, and then compare that to her actual results.

Rather than weigh goals and assists equally by using points, we use the goals created metric. As described in the opening chapter, each goal is worth half a GC and an assist is worth less, depending on how many assists were awarded for each goal. Add up every player's GC, and it should equal the exact number of goals the team scored.

For example, at her peak, Wickenheiser scored five goals and 12 assists in five games at the 2006 Olympics, which works out to 6.7 goals created. That's 2.5 for her five goals and 4.2 for her 12 assists, given the rate at which assists were awarded that year. The average forward on that strong team Canada created 2.9 goals, which means that Wickenheiser created 3.8 goals above expectations (GCAX) during that tournament. As incredibly dominant a performance as that was, she did it again the following season at the 2007 World Championship, where she registered an even higher GCAX of 4.5. As demonstrated in the following table, very few players have performed at that level over the past 20 years.

GOALS CREATED ABOVE EXPECTATIONS IN AN
INDIVIDUAL WORLD CHAMPIONSHIP, 1998 TO 2017[69]

EVENT	PLAYER	TEAM	GP	G	A	GC	TEAM G	GCAX
2010 OG	Stefanie Marty	Switzerland	5	9	2	5.3	14	4.9
2012 WC	Monique Lamoureux	USA	5	7	7	5.9	43	4.6
2007 WC	Hayley Wickenheiser	Canada	5	8	6	6.2	32	4.5
1998 OG	Riikka Valila	Finland	6	7	5	6.1	31	4.2
2001 WC	Yekaterina Pashkevich	Russia	5	6	4	4.9	15	4.2
2006 OG	Hayley Wickenheiser	Canada	5	5	12	6.7	46	3.8
2010 OG	Meghan Agosta	Canada	5	9	6	6.5	48	3.7
2006 OG	Cherie Piper	Canada	5	7	8	6.3	46	3.5

69. World Championship raw data for these calculations from Elite Prospects, accessed May 16, 2017, http://www.eliteprospects.com.

EVENT	PLAYER	TEAM	GP	G	A	GC	TEAM G	GCAX
2001 WC	Cammi Granato	USA	5	7	6	5.9	43	3.4
2007 WC	Krissy Wendell-Pohl	USA	5	5	7	4.9	27	3.3
2017 WC	Kendall Coyne	USA	5	5	7	4.7	28	3.3

The only two players to top Wickenheiser's performance in a single tournament are Switzerland's Stefanie Marty and the USA's Monique Lamoureux. In the former case, Marty scored 9 of the team's 14 goals, so it's easy to see how that could statistically rate among the most dominant single-tournament performances. The latter case is based on Lamoureux being listed as a defenceman, whose scoring expectations are much lower.[70] However, this could be a glitch with the data, since I clearly recall Lamoureux playing forward, and her goal totals are not lower than her assists. That casts more than a little doubt on the argument that her performance rivaled Wickenheiser's.

Add the numbers up for every tournament, and the new GCAX-based leaderboard still has Wickenheiser on top, but she is now followed by someone who is allegedly sometimes on defence, Lamoureux, and players competing for other nations, like Finland's Karoliina Rantamaki and Michelle Karvinen, Sweden's Erika Holst, China's Rui Sun, and Russia's Yekaterina Smolentseva.

GOALS CREATED ABOVE EXPECTATIONS IN WORLD CHAMPIONSHIPS, 1998 TO 2017[71]

PLAYER	TEAM	GP	PTS	GC	EXP	GCAX
Hayley Wickenheiser	Canada	76	127	52.8	31.0	21.7
Monique Lamoureux	USA	43	63	25.3	8.9	16.4
Michelle Karvinen	Finland	50	50	20.9	6.1	14.8
Karoliina Rantamaki	Finland	97	55	22.9	8.9	14.1
Erika Holst	Sweden	61	58	25.4	11.4	14.0

70. Monique Lamoureux was listed as a defenceman at Elite Prospects, accessed May 16, 2017, http://www.eliteprospects.com/team.php?team=19290&year0=2012&status=stats.

71. World Championship raw data for these calculations from Elite Prospects, accessed May 16, 2017, http://www.eliteprospects.com.

PLAYER	TEAM	GP	PTS	GC	EXP	GCAX
Rui Sun	China	46	39	18.4	4.6	13.7
Yekaterina Smolentseva	Russia	70	52	24.0	10.3	13.7
Brianne Decker	USA	35	64	25.3	12.2	13.1
Natalie Darwitz	USA	55	83	35.3	22.2	13.1
Hilary Knight	USA	55	77	32.1	19.7	12.5

It bears repeating that this is all based on scoring data alone. That means that a player's defensive abilities are being ignored, beyond the talent that gets them on these teams in the first place and earns them additional ice time with which to generate scoring.

Frankly, there's just no way around that. There's no way to measure a player's defensive abilities with the available data. In fairness, it is often argued that the highest-scoring players in men's hockey are also the best, whether it's Gordie Howe, Bobby Orr, Wayne Gretzky, Mario Lemieux, Sidney Crosby, or Connor McDavid, so this scoring-based approach is probably not too far off track.

From this point forward, we'll be narrowing our view to just the last three seasons, to create a picture of who is the world's best women's hockey player today. Though the picture may radically change as we add information from other leagues, our early candidates include all three players from our sanity check, with one of them clearly out front. That's a promising start.

GOALS CREATED ABOVE EXPECTATIONS IN WORLD CHAMPIONSHIPS, 2014 TO 2017[72]

PLAYER	TEAM	GP	PTS	GC	EXP	GCAX
Hilary Knight	USA	15	30	13.0	4.7	8.3
Lara Stalder	Switzerland	15	21	8.7	1.9	6.8
Brianna Decker	USA	15	29	11.2	5.2	6.0
Olga Sosina	Russia	16	11	5.3	1.0	4.3
Michelle Karvinen	Finland	16	16	6.7	2.4	4.3
Christine Meier	Switzerland	11	14	5.6	1.4	4.2

---

72. World Championship raw data for these calculations from Elite Prospects, accessed May 16, 2017, http://www.eliteprospects.com.

PLAYER	TEAM	GP	PTS	GC	EXP	GCAX
Monique Lamoureux	USA	15	20	7.3	3.1	4.2
Marie-Philip Poulin	Canada	15	18	7.4	3.3	4.1
Jocelyne Lamoureux	USA	14	20	8.6	4.6	4.0
Jenni Hiirikoski	Finland	18	13	5.4	1.5	3.9
Natalie Spooner	Canada	15	16	6.9	3.2	3.7
Rebecca Johnston	Canada	15	17	6.9	3.4	3.5
Anna Borgqvist	Sweden	14	11	5.1	1.6	3.5
Kendall Coyne	USA	15	22	8.8	5.3	3.5

Translating Data from Other Leagues

Even though the world's best women's hockey player has almost certainly competed in the Olympics and/or the IIHF World Championship, there's just not enough data to evaluate everybody's performance at that level. That's why we need to consider data from professional leagues, universities, and colleges.

However, we can't just add the data from the various leagues together, since each league has different rules, different scoring levels, a different number of players per team, different levels of competition, and so on. Before we combine it, we need to adjust the data for these different factors.

Historically, in baseball this was done as far back as the mid-1980s, and it has been done in hockey since the mid-2000s, and was covered quite extensively in a previous chapter. The process boils down to finding a group of players who have competed in a certain number of games in both leagues, comparing their scoring totals, and using that data to compute the translation factor required to compare a player's data across different leagues.

Let's begin with Wickenheiser's former league, the CWHL, and the closely related NWHL.

Just as men's hockey was spread out across several professional leagues in its early days, like the CAHL, FAHL, ECAHA, NHA, NHL, PCHA, and WCHL, North American professional women's hockey has been spread out across several leagues, the most prominent of which include the Canadian Women's Hockey League (CWHL) and the National Women's Hockey League (NWHL).

There's not enough data to prove it, but it is reasonable to consider the two leagues as being equal. Just like the NHL and PCHA a hundred years ago, the two leagues have very similar rules, structure, and players.

Once adjusted for year-to-year changes in league goals per game, combining data from the CWHL and NWHL is no different from combining those of the original professional men's leagues. That is, it won't be a perfect match, but it's close enough to suit most purposes.

To confirm this, there are 17 players who have played at least 20 games in both leagues, and their average scoring rates in both leagues is exactly 0.73 points per game. While that is normally not enough data on which to base a judgment, that astonishingly exact match makes it reasonable to proceed in treating both leagues equally.

If we add together each player's data in both the CWHL and NWHL and then adjust for the impact of position and quality of teammates (as described on page 156), we get the following leaderboard.

GOALS CREATED ABOVE EXPECTATIONS IN THE CWHL AND NWHL, 2014–15 TO 2016–17[73]

PLAYER	TEAM	GP	G	A	PTS	GC	EXP	GCAX
Brianna Decker	Boston	45	44	48	92	38.1	12.4	25.7
Ann-Sophie Bettez	Montreal	70	48	55	103	40.7	15.9	24.8
Marie-Philip Poulin	Montreal	45	38	45	83	32.8	11.1	21.7
Natalie Spooner	Toronto	62	37	28	65	27.7	7.7	20.0

73. CWHL raw data from the Canadian Women's Hockey League, accessed May 16, 2017, http://www.thecwhl.com, and NWHL raw data from National Women's Hockey League, accessed May 16, 2017, http://www.nwhl.zone.

PLAYER	TEAM	GP	G	A	PTS	GC	EXP	GCAX
Jess Jones	Brampton	72	38	39	77	32.2	12.9	19.3
Caroline Ouellette	Montreal	68	38	51	89	34.4	16.1	18.3
Hilary Knight	Boston	40	31	39	70	28.6	11.4	17.2
Laura Fortino	Brampton	68	19	43	62	24.3	7.5	16.8
Rebecca Johnston	Calgary	48	28	37	65	25.5	11.5	13.9
Jamie Lee Rattray	Brampton	66	28	35	63	26.0	12.5	13.5

Subjectively, it's reassuring to see one of the players from our sanity check in first, Decker, and to also see Poulin in third and Knight in seventh.

Now we need to know in what proportion to combine this data with the Olympic data, since it's a bit more difficult to score in the Olympics and/or World Championship than in the CWHL and NWHL.

There are 38 players who have played at least 20 games in both the Olympics and/or World Championship combined and in the CWHL and NWHL combined. In terms of points per game, there is an average drop from 1.03 in the CWHL and NWHL to 0.82 in the Olympics and World Championship, for a translation factor of 0.80. Statistically, there's an excellent correlation of 0.7 between the two sets of data, which can even be seen in the following chart.

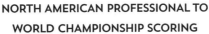

NORTH AMERICAN PROFESSIONAL TO
WORLD CHAMPIONSHIP SCORING

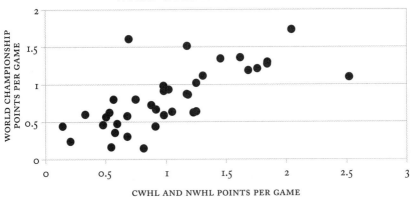

That means that we can add up the data and assume that creating goals in the North American professional leagues is roughly 80% as difficult as doing so in the Olympics and World Championshis. Lo and behold, the three players from our sanity check fall one, two, and three. Now we're cooking with gas! (Note that, due to rounding errors, goals plus assists won't always equal points.)

WEIGHTED SCORING AND GCAX, NORTH AMERICAN PROFESSIONAL LEAGUES AND WORLD CHAMPIONSHIP, 2014–15 TO 2016–17[74]

PLAYER	TEAMS	GP	G	A	PTS	GC	EXP	GCAX
Brianna Decker	USA/Boston	60	45	57	103	41.7	15.1	26.6
Hilary Knight	USA/Boston	55	43	43	86	35.9	13.8	22.1
Marie-Philip Poulin	Canada/Montreal	60	37	47	84	33.7	12.2	21.5
Ann-Sophie Bettez	Canada/Montreal	70	38	44	82	32.5	12.7	19.9
Natalie Spooner	Canada/Toronto	77	38	30	68	29.1	9.4	19.7
Jess Jones	Brampton	72	30	31	62	25.8	10.3	15.4
Caroline Ouellette	Canada/Montreal	73	32	43	75	29.4	14.0	15.4
Rebecca Johnston	Canada/Calgary	63	28	41	69	27.3	12.6	14.7
Laura Fortino	Canada/Brampton	83	17	37	55	21.4	8.0	13.4
Jamie Lee Rattray	Canada/Brampton	72	22	28	50	20.8	11.5	9.3

While we're starting to collect a reasonable amount of data on each player, let's dig deeper and consider the universities and colleges from which these players came before we expand the search into other countries.

U Sports

The best U Sports (the new name of the recently rebranded Canadian Interuniversity Sport, or CIS) player of recent years has clearly been Iya Gavrilova, who played for the University of Calgary Dinos from

74. CWHL raw data from the Canadian Women's Hockey League, accessed May 16, 2017, http://www.thecwhl.com; NWHL raw data from National Women's Hockey League, accessed May 16, 2017, http://www.nwhl.zone; World Championship raw data from Elite Prospects, accessed May 16, 2017, http://www.eliteprospects.com.

2011–12 to 2015–16. If the world's best women's hockey player competed in U Sports recently, then it's probably her.

Gavrilova was the engine behind an otherwise struggling team. In 2015–16, the Dinos scored 64 goals, of which Gavrilova scored 20 and assisted on 23 others. That works out to 17.5 goals created and a GCAX of 14.3 in 28 games. If that seems unusually high, the Dinos fell to a record of 5-19-4 without Gavrilova in 2016–17 and scored just 42 goals in 28 games.

Gavrilova's success was reminiscent of Wickenheiser's time at the University of Calgary. In 2010–11, Wickenheiser earned a GCAX of 13.9 with 40 points in 15 games for the Dinos. That's just shy of the highest single-season GCAX in modern U Sports history, which is held by Mariève Provost, who had a GCAX of 16.2 when she scored 29 goals and 22 assists in 24 games for Moncton in 2009–10.

Adding together every player's totals over the past three seasons, it appears that Gavrilova isn't the only effective U Sports player. Windsor's Krystin Lawrence outscored her 47 goals to 41, and McGill's Mélodie Daoust, who played for Canada in the 2014 Olympics, edged Gavrilova with a GCAX of 0.54 per game compared to Gavrilova's 0.52. Depending on how well that translates to the World Championship level, all three players could theoretically rank among the world's best.

GOALS CREATED ABOVE EXPECTATIONS
IN U SPORTS, 2014-15 TO 2016-17[75]

PLAYER	TEAM	GP	G	A	PTS	GC	EXP	GCAX
Iya Gavrilova	Calgary	51	41	39	80	33.2	6.8	26.4
Krystin Lawrence	Windsor	69	47	38	85	35.9	10.9	25.0
Mélodie Daoust	McGill	41	32	39	71	28.4	6.1	22.2
Alanna Sharman	Manitoba	75	36	45	81	32.7	11.2	21.5
Breanna Lanceleve	Saint Mary's	70	32	55	87	33.2	12.4	20.8
Daley Oddy	St. Francis Xavier	65	33	44	77	31.0	10.5	20.5
Kara Power	St. Francis Xavier	72	34	43	77	31.1	11.9	19.2
Taylor Day	Toronto	65	35	28	63	26.6	8.4	18.3

75. U Sports raw data from U Sports, accessed May 16, 2017, http://en.usports.ca/sports/wice.

PLAYER	TEAM	GP	G	A	PTS	GC	EXP	GCAX
Kaitlin Willoughby	Saskatchewan	78	31	36	67	27.4	9.6	17.8
Lauren Henman	St. Thomas	70	40	29	69	29.5	12.3	17.2

Based on how Gavrilova's success persisted outside U Sports, these results clearly have value. In fact, it would appear that being the top U Sports player equates to being roughly the 20th best player in the world.

Since 2004, Gavrilova has represented Russia at the Olympics and World Championship, at which she has scored 20 goals and 22 assists in 71 games for 18.2 goals created and a career GCAX of 9.7, which ranks as the 18th highest of all time. Joining the CWHL's Calgary Inferno for the 2016–17 season, Gavrilova scored 21 points in 20 games, for 8.8 goals created and a GCAX of 3.2, which ranked 22nd in the league.

While it's clear that U Sports data has enough value to be included in our analysis, it is not possible to translate it directly to the World Championship standard, as we did with the CWHL and NWHL data. Over the last eight years, there have been only seven players who have played at least 20 games in both U Sports and the World Championships, and the results are skewed by the inclusion of exceptional players like Wickenheiser and Gavrilova.

The fact that there are only seven U Sports players who played on World Championship teams is also evidence that the world's best hockey player is unlikely to be found here. Nevertheless, U Sports data can be valuable in filling in some gaps and painting a more accurate picture of women's hockey overall, so it's important to find some other way of including this data.

There may not be enough data to translate U Sports stats directly to a World Championship standard, but there is ample data to translate it to the North American professional standard, which we already know how to adjust to the World Championship standard. This two-step approach is crude—but effective.

From 2009–10 to 2016–17, Gavrilova and Wickenheiser were two of 49 players who played in at least 20 games in U Sports and in either the CWHL and/or NWHL. The group's average scoring rate dropped from 0.88 in U Sports to 0.40 in the North American professional leagues, for an approximate translation factor of 0.45.

Factor in the 0.80 translation factor that exists when going from the North American professional leagues to the World Championship, and U Sports has an estimated two-step translation factor of 0.36.

Unlike the previous chart that compared individual scoring rates from the North American professional leagues to the World Championship, comparing U Sports data to the CWHL and NWHL does not create a clear, straight line (unless you hold the page at arm's length and really squint). This could be because of the varying quality of teams and conferences within U Sports, the relative lack of data, or something else entirely.

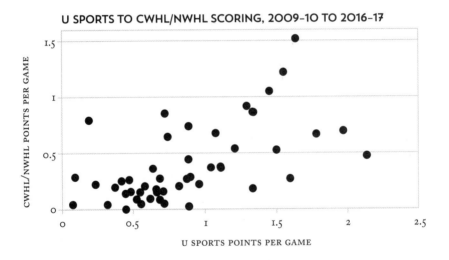

U SPORTS TO CWHL/NWHL SCORING, 2009-10 TO 2016-17

To see a weaker trend more clearly, it can be helpful to bundle several data points together, as we did when translating AHL goaltending data to the NHL. I assembled the following chart in much the same fashion, by combining each point with the two that immediately preceded and succeeded it. Note that there's no statistical validity to these so-called *smoothed* charts and that this chart serves only to make the existing trend a little bit easier to observe.

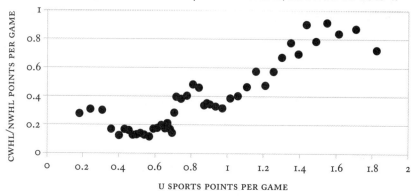

SMOOTHED U SPORTS TO CWHL/NWHL SCORING, 2009-10 TO 2016-17

Even after smoothing, there is still a slight skew at the left of the chart caused by Jessica Campbell and then again at the 0.8 points per game mark caused by Elana Lovell and Devon Skeats, but otherwise we can now see the trend.

At the right side of the chart, the trend starts to taper off at a peak of around 1.5 points per game, beyond which extra scoring doesn't really seem to help predict further success. Wickenheiser, for example, averaged 1.97 points per game in U Sports but 0.7 in her lone CWHL season. She's not alone, as high-scoring U Sports players like Amanda Parkins, Leslie Oles, Katia Clement-Heydra, and Marième Provost didn't exactly light the North American professional leagues on fire either.

In my view, this tapering off of the high-scoring players raises some important points specific to women's hockey that are worth emphasizing, even if they're not directly related to the task at hand.

In men's hockey, the abundance of players means that there are only gradual differences from one line on the depth chart to the next, but that isn't the case in women's hockey. Whether in U Sports or the professional leagues, the top players can be dramatically better than those on the secondary lines, meaning that the latter won't get nearly as much ice time and are unlikely to score very much even when they do.

That's why the average quality of linemates and competition that a player faces generally doesn't vary that significantly from one player to the next in most men's leagues, but it can be downright definitive in women's leagues. Playing with (or against) a top player like Wickenheiser will determine a player's scoring to a far greater extent than how talented that player is.

Let's take one simple example and compare Leslie Oles and Ann-Sophie Bettez, who each played for the McGill Martlets before joining the Montreal Stars/Canadiennes in the CWHL. Bettez scored 37 points in 20 games for the Martlets in 2011–12, followed by 33 points in 23 games for the Stars. Oles scored a comparable 31 points in 18 games for the Martlets in 2014–15, but she then scored just 7 points in 22 games for the Canadiennes.

Why the disparity in CWHL scoring? Presumably, because Bettez played on a line with the incomparable Agosta and Ouellette for the Montreal Stars in a season in which the team scored 105 goals, while Oles clearly did not. More recently, Bettez often plays with Poulin, so even if Bettez is a superior player, this is still a case in which the quality of her teammate has been a far greater factor in determining her CWHL scoring than her own skill.

Even if you're one of the readers who is only interested in the NHL, this is one of the reasons why it's still a good idea to study other leagues. The findings that apply to the NHL, like the relatively limited variance and impact of quality of linemates and competition, don't necessarily apply to other leagues. Studying these concepts in other leagues is a great way to develop a greater understanding of them.

Take PDO as another example, which is the simple sum of shooting and save percentage. In the NHL, it has been observed that PDO has a very strong tendency to approach the league average. The cap-imposed parity makes it so hard for teams to corner the market on certain skills that any abnormally low or high individual or team PDO has almost always proven to be temporary.

But does that rule make any sense in the CWHL or NWHL? While the topic has yet to be explored and is outside the scope of this chapter, it makes perfect sense that a team fortunate enough to have both Knight and Decker on the same line would be able to sustain a high shooting percentage on a significantly more permanent basis. Even in men's hockey, it makes sense that the Canadian national team would sustain a PDO that was much higher than the Latvian team.

Unfortunately, the realities of the available data for women's hockey can be disruptive in our quest for the world's best player. Since we don't know how much ice time a player gets, let alone how much of it is spent with or without certain players, it's impossible to place any player's scoring in an appropriate context. It also explains why a player's scoring can be so radically different in the CWHL or NWHL and in U Sports.

Using stats like GCAX does help, to an extent. It doesn't help adjust for which line a player is on, but it does help account for the strength of the team in general. For example, Tatiana Rafter scored 12 points in 23 games for the UBC Thunderbirds in 2011–12. Two seasons later, she scored 38 points in 28 games for the same team and was named Canada West Player of the Year. While that appears to be a big personal improvement, it probably wasn't.

UBC scored 22 goals in 24 games en route to a 1-21-2 record in 2011–12, and they then scored 81 goals in 28 games for a 20-6-2 record two years later, in 2013–14. Rafter was in on 54.5% of the team's scoring in the former season and on 46.9% of their scoring in the latter, suggesting that she always would have been among the league's leading scorers had UBC been a stronger team. GCAX adjusts for this particular situation quite nicely.

Another key challenge with U Sports women's hockey is the quality of the data. For example, it appears rather impressive that Jenna Currie scored 9 goals in 12 games for the Dalhousie Tigers in 2012–13, who scored just 14 goals in the 24-game season as a team. However, that year the team had to forfeit half the season because 19 of the 24 players were suspended for a hazing incident. In reality, the Tigers scored 14 goals in 12 games and then 0 goals in the 12 games that were forfeited. Using the raw data without knowing details like that will further skew the results for various players and teams.

While these are all key points when studying women's hockey statistics, they don't really get in the way of our mission, since the world's best women's hockey player is rarely still in U Sports. So other than adding the best recent U Sports player to the second-last position of the list, the U Sports data doesn't really change our leaderboard.

PLAYER	TEAMS	GP	G	A	PTS	GC	EXP	GCAX
Brianna Decker	USA/Boston	60	45	57	103	41.7	15.1	26.6
Hilary Knight	USA/Boston	55	43	43	86	35.9	13.8	22.1
Marie-Philip Poulin	Canada/Montreal	60	37	47	84	33.7	12.2	21.5
Ann-Sophie Bettez	Canada/Montreal	70	38	44	82	32.5	12.7	19.9
Natalie Spooner	Canada/Toronto	77	38	30	68	29.1	9.4	19.7
Jess Jones	Brampton	72	30	31	62	25.8	10.3	15.4
Caroline Ouellette	Canada/Montreal	73	32	43	75	29.4	14.0	15.4
Rebecca Johnston	Canada/Calgary	63	28	41	69	27.3	12.6	14.7
Iya Gavrilova	Russia/Calgary	88	29	25	54	22.7	8.1	14.5
Laura Fortino	Canada/Brampton	83	17	37	55	21.4	8.0	13.4

U.S. College Hockey

Other than the North American professional leagues, U.S. colleges could be the source of most of the world's best hockey players. Consider the 2017 World Championship, where Kendall Coyne tied Decker, the tournament MVP, for the scoring lead with 12 points in five games and Switzerland's Lara Stalder tied Knight for third, with 9 points. As for Poulin, she recently played at Boston University.

There's also a historical precedent for elite play in the NCAA. Going back to the early 2000s, the best single-season performance was the legendary Jennifer Botterill, who scored 47 goals and 65 assists in 32 games for Harvard in 2002–03. That works out to an amazing 44.5 goals created total and 33.1 goals above expectations, which is more than one extra goal created per game. I'm not sure anyone will ever approach that.

Two years later, in 2004–05, Nicole Corriero was Harvard's star,

76. CWHL raw data from Canadian Women's Hockey League, accessed May 16, 2017http://www.thecwhl. com; NWHL raw data from National Women's Hockey League, accessed May 16, 2017, http://www.nwhl. zone; World Championship raw data from Elite Prospects, accessed May 16, 2017, http://www.eliteprospects. com; U Sports raw data from U Sports, accessed May 16, 2017, http://en.usports.ca/sports/wice.

scoring 59 goals in 36 games. However, she may have been upstaged by Natalie Darwitz and Krissy Wendell, who formed a great pair for the University of Minnesota and combined for 85 goals in 40 games. Wowsers.

Fast forward to the present, and Coyne tops the leaderboard over the past three seasons combined, with Stalder in seventh and some very high-scoring players in between, especially at Boston College.

GOALS CREATED ABOVE EXPECTATIONS IN THE NCAA DIVISION I, 2014–15 TO 2016–17[77]

PLAYER	TEAM	GP	G	A	PTS	GC	EXP	GCAX
Kendall Coyne	Northeastern	68	78	58	136	57.1	11.9	45.3
Alex Carpenter	Boston College	78	80	89	169	65.9	22.7	43.2
Cayley Mercer	Clarkson	117	75	77	152	60.9	23.3	37.6
Dani Cameranesi	Minnesota	101	74	91	165	63.8	26.7	37.1
Annie Pankowski	Wisconsin	115	68	88	156	60.2	26.0	34.2
Haley Skarupa	Boston College	78	66	84	150	57.4	23.6	33.7
Lara Stalder	Minnesota Duluth	106	50	76	126	48.9	16.9	32.0
Denisa Krizova	Northeastern	104	51	88	139	52.7	21.2	31.5
Hannah Brandt	Minnesota	76	59	79	138	52.6	21.3	31.3
Ashleigh Brykaliuk	Minnesota Duluth	110	51	71	122	48.0	17.5	30.4

Unlike U Sports data, these results can easily be added with great confidence to our growing collection of data, and without the same two-step approach. The 68 players who have played at least 20 games in both NCAA hockey and the World Championship have an average scoring rate of 1.15 points per game in the NCAA and 0.69 in the World Championship, which works out to a translation factor of 0.6, which is clear in the chart below. There may even be greater distinctions if the data is divided into the various conferences, like ECAC and Hockey East, but let's tag that as future work.

---

77. U.S. college hockey raw data from U.S. College Hockey Online, accessed May 16, 2017, http://www .uscho.com/d1women/, collected with the assistance of Derek Braid.

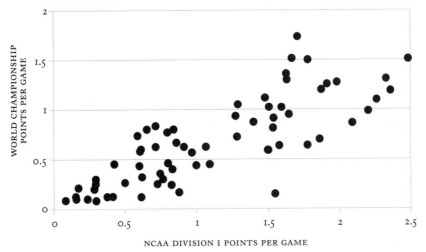

NCAA DIV I TO WORLD CHAMPIONSHIPS, 2000-01 TO 2016-17

NCAA DIVISION I POINTS PER GAME

In addition to these 68 data points, there are also 179 players who have played at least 20 games in both the NCAA and either the CWHL or NWHL. Their average scoring rate dropped from 0.79 in the NCAA to 0.55 points per game in the pro leagues, which results in a translation factor of 0.7.

That translation factor of 0.7 is actually just a touch lower than expected. Given that the NCAA's World Championship translation factor is 0.6 and the CWHL and NWHL's is 0.8, the translation factor between the two sets of leagues should be the former divided by the latter, which is 0.75. That's still within the bounds of reason for such rough estimates, but it does signal the possibility that the NCAA's actual translation factor is a little bit lower than calculated (i.e., lower than 0.6).

So far, we're making a distinction between Division I and Division III NCAA hockey. To this point, the data has included only the former. In men's hockey, the analysis of U.S. college hockey actually ends with Division I, since there's really no chance of those in the other divisions of ever playing in the professional leagues. That's not the case in women's hockey, where Division III can be quite competitive, on a par with some of the European leagues we'll be looking at in the next section, and they can produce a few players

-THE FUTURE of JERSEY ADS-

who are capable of playing in the World Championship and/or the North American professional leagues.

Unfortunately, there are too few such players to actually nail down an accurate translation factor for Division III. Only two NCAA Division III players have met the 20-game threshold at the World Championship, Austria's Monika and Nina Waidacher. As for the two-step approach, only 24 have played at least 20 games in a North American professional league, and even a smoothed version of that translation doesn't paint a particularly clear picture.

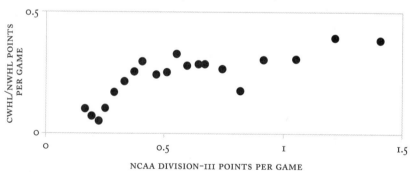

SMOOTHED DIV-III TO CWHL/NWHL SCORING, 2009-10 TO 2016-17

For what it's worth, these players averaged 0.64 points per game in Division III and 0.24 points per game in the North American professional leagues, which results in a translation factor of 0.38. Once multiplied by the CWHL and NWHL translation factor of

0.8, that yields a two-step World Championship translation factor of roughly—very roughly—0.3 for Division III hockey.

While it is unlikely to affect our quest for the world's best women's hockey player, it doesn't hurt to include this data. Based on GCAX, the most effective player in Division III hockey is Dani Sibley of Wisconsin–River Falls. In 2016–17, she scored 27 goals and 38 assists in 29 games, for a GCAX of 19.7. That ranks sixth since the year 2000, not far behind Melanie Salatino, who scored 36 goals and 25 assists in 26 games for Wisconsin Superior in 2001–02, which works out to 23.5. However, adding this data only boosts Sibley up to 41st on the overall leaderboard.

GOALS CREATED ABOVE EXPECTATIONS IN THE NCAA DIVISION III, 2014-15 TO 2016-17[78]

PLAYER	TEAM	GP	G	A	PTS	GC	EXP	GCAX
Dani Sibley	Wisconsin–River Falls	87	65	77	142	60.2	21.9	38.4
Kristin Lewicki	Adrian	83	74	74	148	60.1	22.8	37.3
Melissa Sheeran	Plattsburgh	77	74	52	126	54.4	22.4	32.0
Kayla Meneghin	Plattsburgh	84	62	79	141	57.5	25.5	32.0
Kaylyn Schroka	Adrian	83	62	67	129	52.2	23.7	28.5
Carly Moran	Wisconsin–River Falls	88	57	61	118	50.5	22.8	27.7
Chelsea Blackburn	Stevenson	77	63	28	91	42.3	15.0	27.3
Bek Lucas	Neumann	67	45	29	74	34.0	7.3	26.7
Chloe Kinsel	Wisconsin–River Falls	59	46	47	93	40.7	14.7	25.9
Kathryn Larson	Bethel	106	58	48	106	44.6	19.6	25.0

On the other hand, adding the Division I data to our analysis has a very noticeable impact on the results. For starters, it thrusts Poulin into the top spot, due to her success at Boston University, and it pushes Decker and Knight down the list to fourth and ninth, albeit in far fewer games. On a per-game basis, Decker is still in first with 0.44, followed by Knight with 0.40, then Coyne at 0.37, and Poulin at 0.33.

---

78. U.S. college hockey raw data from U.S. College Hockey Online, accessed May 16, 2017, http://www.uscho.com/dIwomen/, collected with the assistance of Derek Braid.

PLAYER	TEAMS	GP	G	A	PTS	GC	EXP	GCAX
Marie-Philip Poulin	Canada/Montreal/ Boston Univ.	92	54	63	117	47.0	16.4	30.7
Kendall Coyne	USA/Northeastern Univ.	83	56	48	104	43.0	12.4	30.7
Alex Carpenter	USA/Boston/ Boston Coll.	110	59	72	132	51.3	23.4	27.9
Brianna Decker	USA/Boston	60	45	57	103	41.7	15.1	26.6
Lara Stalder	Switzerland/Univ. of Minnesota Duluth	121	38	59	97	38.1	12.1	26.0
Cayley Mercer	Clarkson	117	45	46	91	36.5	14.0	22.5
Haley Skarupa	USA/Connecticut/ Boston Coll.	109	51	65	117	45.1	22.8	22.3
Dani Cameranesi	Univ. of Minnesota	101	44	55	99	38.3	16.0	22.2
Hilary Knight	USA/Boston	55	43	43	86	35.9	13.8	22.1
Ann-Sophie Bettez	Montreal	70	38	44	82	32.5	12.7	19.9

U.S. college hockey has added two new players to the leaderboard who have never competed in a World Championship, Cayley Mercer and Dani Cameranesi. At the moment, they appear to still be a step back of the big names, but if they do break out as stars on Team USA in the future, then remember where you read it first.

As for Stalder, she's starting to look lonely as the only European on the list, so let's cross the Atlantic to complete that part of the picture.

---

79. CWHL raw data from the Canadian Women's Hockey League, accessed May 16, 2017, http://www.thecwhl.com; NWHL raw data from the National Women's Hockey League, accessed May 16, 2017, http://www.nwhl.zone; World Championship raw data from Elite Prospects, accessed May 16, 2017, http://www.eliteprospects.com, U Sports raw data from U Sports, accessed May 16, 2017, http://en.usports.ca/sports/wice; NCAA raw data from U.S. College Hockey Online, accessed May 16, 2017, http://www.uscho.com/d1women/, with the assistance of Derek Braid.

In men's hockey, the most competitive league outside of the NHL is Russia's KHL. Legends like Pavel Datsyuk and Ilya Kovalchuk play there, and KHL players who go to the NHL tend to immediately perform well, like Alexander Radulov, Artemi Panarin, and Nikita Zaitsev. However, this is not the case in women's hockey.

The average scoring rate of the 18 players who competed in at least 20 games in both the top Russian league over the past two seasons and the World Championship over those same years dropped from 0.91 to 0.26, which works out to a translation factor just below 0.30. That places the Russian league on a par with the NCAA's Division III and, therefore, unlikely to include the world's best players.

Is the competition level of the top Russian league really that mediocre? It's hard to be sure. The league is pretty small, reliable data can be very hard to find beyond the last two seasons, not many of these players have competed in other leagues, and the translation chart is all over the place, even when smoothed.

There is certainly some anecdotal evidence that the competition level of the Russian league may be as modest as we think. For example, Lyudmila Belyakova scored 51 points in 33 games for Tornado Dmitrov, which is 1.55 points per game. In her lone NWHL season, she scored 10 points in 15 games, which is 0.67 points per game.

Or consider Alena Polenska, who scored 1.09 points per game in the Russian league. At Brown University, which is a Division I school, she scored 0.62 points per game. These two examples peg the Russian league at around a 0.35 translation factor, so it does seem reasonable to classify it as more of a secondary league.

Of course, that's no reason to exclude its data, especially since we gladly included NCAA Division III data. The top player is arguably Anna Shokhina, who scored 33 goals and 36 assists in 33 games for Tornado Dmitrov in 2016–17, which works out to 29.8 goals created and 20.7 goals created above expectations. Olga Sosina, who has been highly effective in World Championships, ranks second.

GOALS CREATED ABOVE EXPECTATIONS IN
TOP RUSSIAN LEAGUE, 2015–16 TO 2016–17[80]

PLAYER	TEAM	GP	G	A	PTS	GC	EXP	GCAX
Anna Shokhina	Tornado Dmitrov	57	53	67	120	51.2	15.1	36.0
Olga Sosina	Agidel Ufa	52	47	62	109	47.3	15.2	32.1
Fanuza Kadirova	Arktik-Universitet Ukhta	54	35	42	77	36.9	9.1	27.8
Yelena Dergachyova	Tornado Dmitrov	54	31	71	102	41.6	15.8	25.9
Karoliina Rantamaki	SKIF Nizhny Novgorod	52	31	31	62	30.5	5.4	25.1
Alexandra Huszak	Arktik-Universitet Ukhta	49	33	32	65	31.0	8.5	22.5
Valeria Pavlova	Biryusa Krasnoyarsk	24	26	17	43	22.4	4.9	17.5
Alevtina Shtaryova	Tornado Dmitrov	45	40	20	60	27.4	12.1	15.2
Alnea Polenska	Dynamo St. Petersburg	57	31	31	62	27.3	12.7	14.5
Lyudmila Belyakova	Tornado Dmitrov	33	31	20	51	22.9	9.2	13.6

So if the top Europeans aren't playing in Russia, where are they? They go to Sweden's SDHL. Not only does the SDHL include Sweden's top players, but it attracts the best from elsewhere in Europe, like Finland's Michelle Karvinen and Austria's Denise Altmann.

The SDHL isn't a big or deep league, but it has a handful of elite players on the top lines. The best single-season performance was from Andrea Dalen, who scored 47 goals and 26 assists in 36 games for Djurgardens in 2015–16, which works out to 32.7 goals created and 27.1 goals created above expectations. The second- and fifth-best single-season performances were by Karvinen, the third-, sixth-, eighth-, and tenth-best seasons were by Altmann, and Canada's Jennifer Wakefield has the fourth- and ninth-best seasons. All four players are featured prominently in the following SDHL leaderboard, and they will likely appear on the world leaderboard once we add their data at the end of this section.

80. Russian league raw data from Elite Prospects, accessed May 16, 2017, http://www.eliteprospects.com.

GOALS CREATED ABOVE EXPECTATIONS IN SDHL, 2014–15 TO 2016–17[81]

PLAYER	TEAM	GP	G	A	PTS	GC	EXP	GCAX
Denise Altmann	Linkoping	99	80	91	171	72.6	25.5	47.1
Jennifer Wakefield	Linkoping	63	90	40	130	59.6	15.1	44.6
Michelle Karvinen	Lulea	67	67	82	149	60.9	16.6	44.3
Lisa Johansson	Vaxjo	96	70	40	110	49.7	14.6	35.1
Andrea Dalen	Linkoping	64	62	38	100	44.8	10.0	34.8
Emma Nordin	MODO	90	60	75	135	54.9	22.6	32.3
Emma Eliasson	Brynas	97	34	83	117	44.8	14.1	30.7
Fanny Rask	Linkoping	98	51	47	98	43.6	15.3	28.3
Erika Grahm	MODO	93	43	54	97	39.3	13.5	25.8
Pernilla Winberg	AIK	83	43	78	121	49.1	23.6	25.5

How strong is the SDHL? There are 38 players who competed in at least 20 games in both the SDHL and the World Championship over the past decade or so. Their average scoring rate dropped from 1.03 points per game in the SDHL to 0.47 points in the World Championship, which works out to a translation factor of 0.45. On a chart, the trend is about as visible as for the NCAA, which had a translation factor of 0.6.

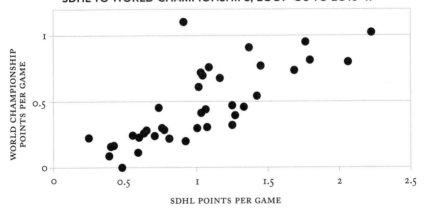

SDHL TO WORLD CHAMPIONSHIPS, 2007–08 TO 2016–17

81. SDHL raw data from Elite Prospects, accessed May 16, 2017, http://www.eliteprospects.com.

Does it make sense that Europe's top league has a translation factor that is 0.15 lower than U.S. college hockey? Yes and no.

Firstly, the points on these charts aren't independent. These players aren't being randomly distributed across the teams competing at the World Championship. Most of the SDHL players are playing for Sweden, or perhaps Finland or Austria, while most of the players in the NCAA are playing for powerhouses like Canada or the USA. That makes it easier for the weaker SDHL players to make their national team but harder to have their scoring boosted by elite linemates. In theory, that could deflate the translation factor.

Imagine if Kendall Coyne played for Team Sweden. Without linemates like Decker and Knight, her scoring in the World Championship would have been lower, and she would not have led the tournament in scoring, right? Now imagine that most of the NCAA's top players played for Team Sweden, and you can see how the translation factor will get pulled down on the whole. Conversely, how would the SDHL's translation factors rise if more of their players played for the USA or Canada?

To put it in NHL terms, what would happen if most of the NCAA players went to particularly good teams, like Chicago and Pittsburgh? Or exclusively to less competitive teams, like Arizona and Buffalo? That bias would probably skew the translation factors in one direction or another.

In reality, the SDHL is probably a bit better than its 0.45 translation factor suggests, and the NCAA is probably a bit worse than its 0.6 translation factor suggests (which we already suspected). If there were more players who played on both sides of the Atlantic, then we could probably calculate how these translation factors should change, or we could try to calculate how much more an identical player would score on Team Canada rather than Team Sweden and make an adjustment.

This phenomenon should be even more pronounced for the Swiss league. The SWHL has an apparent translation factor of 0.18, based on the 21 players who have competed in at least 20 games in the SWHL and the World Championship, whose average points per game dropped from 1.30 to 0.23. Even with so few players and such a low translation factor, you can still see the relationship in a smoothed version of the chart.

SMOOTHED SWHL TO WORLD CHAMPIONSHIPS, 2007–08 TO 2016–17

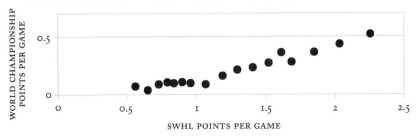

However, how much of these modest World Championship scoring rates were caused because SWHL players primarily played for less competitive teams, like Switzerland?

Let's put our theory to a little test. There are 10 players who have competed in at least 20 games in both the SWHL and NCAA. We have calculated the latter's translation factor to be 0.6. So if the SWHL's translation factor is truly 0.18, then the translation factor between the SWHL and the NCAA should be one divided by the other, which is around 0.3. If there is a bias against those who play for weaker World Championship teams, like in the SWHL, compared to those who play for the powerhouse North American teams, like in the NCAA, then the translation factor between those two leagues will be greater than 0.3. Potentially, it will be far greater.

The results? The 10 players in question had an average scoring rate of 1.31 points per game in the SWHL and 0.38 points per game in the NCAA, which works out to a translation factor of almost precisely 0.3. So the 0.18 translation factor for the SWHL is likely accurate, and this whole digression might have been much ado about nothing, but at least it clarifies things.

While we're here, let's add all the SWHL data to our study. The top players right now are undeniably Christine Meier and Stefanie Marty, along with Nicole Bullo on defence. Meier had the best single-season performance in 2013–14, with 51 goals and 28 assists in 20 games, for 36.4 goals created and 26.5 goals created above expectations. With the exception of Isabel Menard, who had a GCAX of

21.1 that season, nobody has come within 10 goals of that mark in the past decade.

GOALS CREATED ABOVE EXPECTATIONS
IN SWHL, 2014-15 TO 2016-17[82]

PLAYER	TEAM	GP	G	A	PTS	GC	EXP	GCAX
Christine Meier	ZSC Lions	49	74	57	131	59.9	25.8	34.2
Stefanie Marty	Neuchâtel	45	52	46	98	46.4	15.5	31.0
Evelina Raselli	Lugano	56	54	54	108	52.3	27.1	25.2
Isabel Waidacher	ZSC Lions	43	50	56	106	47.3	24.7	22.7
Nina Waidacher	ZSC Lions	35	42	40	82	36.8	20.8	16.0
Nicole Bullo	Lugano	48	23	36	59	28.3	14.0	14.3
Becca Kohler	Lugano	19	31	16	47	23.3	9.0	14.3
Isabel Menard	ZSC Lions	20	19	32	51	23.1	9.4	13.7
Breehan Polci	Neuchâtel	40	36	23	59	28.0	15.3	12.7
Sarah Forster	Neuchâtel	49	12	46	58	25.9	13.4	12.7

Next, there's Finland. A lot of its top players actually play in Sweden's SDHL. In Finland's SM-sarja, the best single season is by Karoliina Rantamaki on defence, who is essentially that country's Hayley Wickenheiser, but who peaked a little sooner. In 2002–03, Rantamaki scored 39 goals and 31 assists in 24 games for the Blues, which works out to 30.9 goals created and 25.0 goals created above expectations. The next season, she had a GCAX of 24.7, which ranks as the third-best single season of all time.

More recently, the high point was set by JYP defenceman Jenni Hiirikoski, who scored 17 goals and 62 assists in 28 games, which works out to 28.7 goals created and 21.0 goals created above expectations. She ranks second on the following SM-sarja leaderboard, behind Linda Valimaki, who has consistently had a GCAX between 17.1 and 18.7 every season from 2012–13 to 2016–17. If Finland is known for its goalies in the men's game, then it's known for its defenders in the women's game.

---

82. SWHL raw data from Elite Prospects, accessed May 16, 2017, http://www.eliteprospects.com.

GOALS CREATED ABOVE EXPECTATIONS
IN SM-SARJA, 2014–15 TO 2016–17[83]

PLAYER	TEAM	GP	G	A	PTS	GC	EXP	GCAX
Linda Valimaki	Espoo United	84	88	107	195	77.9	25.4	52.6
Jenni Hiirikoski	JYP	56	35	105	140	51.1	14.1	36.9
Matilda Nilsson	KalPa	81	62	32	94	43.6	11.0	32.5
Johanna Juutilainen	KalPa	81	38	59	97	42.6	13.1	29.4
Minnamari Tuominen	Blues	54	30	77	107	39.0	11.3	27.7
Riikka Noronen	HPK	81	40	81	121	47.4	21.3	26.1
Annina Rajahuhta	Espoo United	64	50	68	118	46.7	20.9	25.8
Sari Karna	Ilves	75	52	61	113	46.3	21.6	24.7
Tanja Niskanen	KalPa	54	51	47	98	41.9	17.7	24.2
Saila Saari	Karpat	84	55	79	134	51.9	30.8	21.2

Finland's translation factor is 0.3, based on the 37 players who played at least 20 games in both the SM-sarja and the World Championship since the year 2000. Their average points per game dropped from 1.17 to 0.35. Those stats include games against Team China, who played as a "guest" team in the 2005–06 and 2006–07 seasons. That places Finland roughly on a par with the Russian league and with Division III in the NCAA. It's not a perfect mapping though, even with the smoothed version of the chart.

SMOOTHED SM-SARJA TO WORLD CHAMPIONSHIPS,
2000–01 TO 2016–17

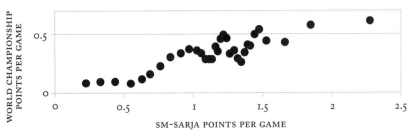

For the sake of completeness, we'll close with the European Women's Hockey League (EWHL), which isn't even at the same

83. SM-sarja raw data from Elite Prospects, accessed May 16, 2017, http://www.eliteprospects.com.

competition level as the SWHL. It's a small league, with very few players who have played in the World Championship and/or the other top leagues, and it has been somewhat dominated by the Vienna-based team.

GOALS CREATED ABOVE EXPECTATIONS IN EWHL, 2014–15 TO 2016–17[84]

PLAYER	TEAM	GP	G	A	PTS	GC	EXP	GCAX
Anna Meixner	Vienna	36	60	42	102	48.2	18.2	30.0
Esther Kantor	Vienna	53	56	71	127	57.7	28.1	29.6
Chelsea Furlani	Flyers	42	41	34	75	33.9	10.8	23.0
Hillary Crowe	Salzburg	34	34	27	61	27.9	7.0	21.0
Charlotte Wittich	Flyers	53	42	29	71	33.5	14.7	18.8
Sarah Campbell	Salzburg	34	27	27	54	24.0	7.6	16.4
Reka Dabasi	Budapest	37	25	33	58	25.0	11.7	13.3
Kaitlin Spurling	Vienna	37	39	32	71	33.4	20.6	12.8
Alex Gowie	Budapest	39	27	16	43	19.6	6.8	12.7
Regan Boulton	Vienna	20	12	30	42	19.2	6.8	12.4

The translation factor for the EWHL is virtually impossible to compute. Very few of these players have competed in any other league. Only four have played at least 20 games in the EWHL and the World Championship, and it was usually for one of the weaker nations that are lucky to average one goal per game. However, there have been 18 players who played at least 20 games in both the EWHL and the SDHL over the past decade or so. Their average scoring dropped from 0.88 points per game in the EWHL to 0.32 in the SDHL, for a translation factor of 0.36. Multiply that by the SDHL's World Championship translation factor of 0.45, and there's a two-step translation factor of 0.16.

Once we add up all of the data from these five leagues, three more European-based players join Stalder among the top dozen players. Stalder is fifth, Karvinen is sixth, Wakefield is seventh, and Altmann ranks 12th.

The addition of European data doesn't otherwise change those

---

84. EWHL raw data from Elite Prospects, accessed May 16, 2017, http://www.eliteprospects.com.

already on the leaderboard, since very few players compete on both sides of the Atlantic. At last, here are the 12 players in the world who have created at least 20 goals over and above expectations by World Championship standards.

WEIGHTED SCORING AND GCAX, ALL WORLD DATA, 2014–15 TO 2016–17[85]

PLAYER	TEAMS	GP	G	A	PTS	GC	EXP	GCAX
Kendall Coyne	USA/Northeastern Univ.	83	56	48	104	43.0	12.4	30.7
Marie-Philip Poulin	Canada/Montreal/ Boston Univ.	92	54	63	117	47.0	16.4	30.7
Alex Carpenter	USA/Boston/ Boston Coll.	110	59	72	132	51.3	23.4	27.9
Brianna Decker	USA/Boston	60	45	57	102	41.7	15.1	26.6
Lara Stalder	Switzerland/Univ. of Minnesota Duluth	121	38	59	97	38.1	12.1	26.0
Michelle Karvinen	Finland/Lulea	83	37	46	83	34.1	9.9	24.2
Jennifer Wakefield	Canada/Linkoping	78	48	26	74	33.3	10.1	23.2
Cayley Mercer	Clarkson Univ.	117	45	46	91	36.5	14.0	22.5
Haley Skarupa	USA/Connecticut/ Boston Coll.	109	51	65	116	45.1	22.8	22.3
Dani Cameranesi	Univ. of Minnesota	101	44	55	99	38.3	16.0	22.2
Hilary Knight	USA/Boston	55	43	43	86	35.9	13.8	22.1
Denise Altmann	Linkoping	99	36	41	77	32.7	11.5	21.1

There are many more women's leagues at the amateur and professional levels throughout Europe, but even the inclusion of the EWHL and the SWHL is starting to push it. The odds that even one of the world's top-50 players is outside of the dozen leagues that we've already covered are pretty long.

---

85. CWHL raw data from Canadian Women's Hockey League, accessed May 16, 2017, http://www .thecwhl.com; NWHL data from National Women's Hockey League, accessed May 16, 2017, http://www .nwhl.zone; raw data for World Championship, EWHL, Russia, SDHL, SWHL, and EWHL from Elite Prospects, accessed May 16, 2017, http://www.eliteprospects.com; U Sports raw data from U Sports, accessed May 16, 2017, http://en.usports.ca/sports/wice; NCAA raw data from U.S. College Hockey Online, accessed May 16, 2017, http://www.uscho.com/d1women/, with the assistance of Derek Braid.

That means that we're ready to pick Wickenheiser's successor. Drum roll, please.

Closing Thoughts

Wickenheiser exploded onto the scene in the mid-1990s as the first mainstream superstar in women's hockey, and she retired in a world with an abundance of exceptional female talent. With such a wide variety of leagues and such limited sets of data, it isn't easy to pick the world's new number one player, but it is possible to start painting a rough picture.

By crafting a version of the goals created metric mentioned in the opening chapter as a slight improvement over points and then measuring it relative to a team's average player to account for the team's overall strength, we can identify the top offensive players in each league. Next, by calculating the difference in scoring rates among players who have seen significant action in different leagues, it's possible to start adding together all of this data in the following measures:

WOMEN'S HOCKEY TRANSLATION FACTORS, AS OF 2016–17

LEAGUE	FACTOR
Winter Olympics	1.00
IIHF World Championship	1.00
Canadian Women's Hockey League (CWHL)	0.80
National Women's Hockey League (NWHL)	0.80
NCAA Division I	0.60
Swedish Women's Hockey League (SDHL)	0.45
U Sports	0.36
Finland's SM-sarja	0.30
NCAA Division III	0.30
Russia's Women's Hockey League	0.30
Swiss Women's Hockey League (SWHL)	0.18
Elite Women's Hockey League (EWHL)	0.16

All of this work brings us to this final leaderboard, which is ordered by the number of goals a player has created above team-based expectations on a per-game basis. From this perspective, Decker ranks first with an extra 0.44 goals created per game, followed by three other players who add over 0.3.

WEIGHTED SCORING AND GCAX, ALL WORLD DATA, 2014–15 TO 2016–17[86]

PLAYER	TEAMS	GP	G	A	PTS	GC	EXP	GCAX	GCAX/GP
Brianna Decker	USA/Boston	60	45	57	102	41.7	15.1	26.6	0.44
Hilary Knight	USA/Boston	55	43	43	86	35.9	13.8	22.1	0.40
Kendall Coyne	USA/ Northeast Univ.	83	56	48	104	43.0	12.4	30.7	0.37
Marie-Philip Poulin	Canada/ Montreal	92	54	63	117	47.0	16.4	30.7	0.33
Jennifer Wakefield	Canada/ Linkoping	78	48	26	74	33.3	10.1	23.2	0.30
Michelle Karvinen	Finland/ Lulea	83	37	46	83	34.1	9.9	24.2	0.29
Ann-Sophie Bettez	Canada/ Montreal	70	38	44	82	32.5	12.7	19.9	0.28
Natalie Spooner	Canada/ Toronto	77	38	30	68	29.1	9.4	19.7	0.26
Alex Carpenter	USA/Bos/ Boston Coll.	110	59	72	131	51.3	23.4	27.9	0.25
Rebecca Johnston	Canada/ Calgary	63	28	41	69	27.3	12.6	14.7	0.23

Minimum 50 games played.

Before we began this exercise, we subjectively chose Decker, Knight, and Poulin as the world's best players. As a sanity check,

86. CWHL raw data from the Canadian Women's Hockey League, accessed May 16, 2017, http://www .thecwhl.com; NWHL raw data from the National Women's Hockey League, accessed May 16, 2017, http:// www.nwhl.zone; raw data for the World Championship, EWHL, Russia, SDHL, SWHL, and EWHL from Elite Prospects, accessed May 16, 2017, http://www.eliteprospects.com; U Sports raw data from U Sports, accessed May 16, 2017, http://en.usports.ca/sports/wice; NCAA raw data from U.S. College Hockey Online, accessed May 16, 2017, http://www.uscho.com/d1women/, with the assistance of Derek Braid.

their inclusion in the top four spots provides some credibility to the process that generated these results.

Coyne's inclusion among them may have also inadvertently helped us identify the world's most underrated player. Then again, a case could also be made for Bettez, who wasn't even included on the most recent Canadian national team.

Unfortunately, lack of data means that players like Jocelyne Lamoureux and Amanda Kessel couldn't be included in this final analysis. Similarly, Switzerland's 19-year-old Alina Muller is too young to have her performance level accurately measured. As such, there remains the possibility that the world's best hockey player slipped through this net.

That's why we should consider this a first stride in a far greater journey. We need to find better and more complete ways to measure a female player's contributions, find more accurate methods to measure and adjust for the impact of the quality of her teams, linemates, and opponents, and refine the process of comparing and/or combining data from across the women's leagues.

Just as there were few women's hockey players when Wickenheiser first laced up her skates, there are few statisticians who are studying women's hockey today. Over the next decade, I hope our own field explodes at the same rate as Wickenheiser's has. Until then, the data points very strongly to the popular subjective view that Decker, Knight, Coyne, and Poulin sit atop a tight pack in a close competition to succeed Wickenheiser as the world's best women's hockey player.

WHO HAS THE BEST
COACHING STAFF?

When I write, I always start off by trying to pull in the reader by illustrating the purpose of the upcoming analysis, so that it's clear why the subject material is both interesting and important enough to warrant both my and the reader's effort. In this case, I'm just going to state it flat out: a team's coaching staff is extremely important, and we do a lousy job at evaluating them subjectively.

There is no better example of this phenomenon than the Jack Adams Award, which the NHL Broadcasters' Association has awarded to the league's best coach since 1973–74. In theory, it is meant to celebrate the incredible achievements of great coaches like Scotty Bowman and Al Arbour. In reality, it hasn't proven to be very useful at identifying the best coaches. Firstly, even Bowman and Arbour have only won the award twice in 24 years of eligibility and once in 19 years of eligibility, respectively. It seems strange for two legendary coaches who combined for 12 Stanley Cups since 1973–74 to have a combined total of just three Jack Adams Awards. Of course, there are few coaches of whom they can be jealous, since only four other coaches have won it more than once, and only Pat Burns has won it three times.

The Jack Adams often seems to go to the luckiest coach, instead of the best. At times, it seems that the broadcasters are basing their vote on which teams they held in the lowest regard but who managed to be competitive anyway. Since Bowman and Arbour were usually behind the benches of strong teams that were expected to do well, they were rarely recognized as being the league's best coaches.

To take a more recent example, consider the 2012–13 season, in which Paul MacLean of the Ottawa Senators was awarded the Jack Adams instead of Joel Quenneville of the Chicago Blackhawks. Remember that was the year that the Blackhawks lost just seven games in regulation all season, for a 36-7-5 record and an 0.802 winning percentage that stands as the fifth best of all time and the best since Bowman's own 1977–78 Montreal Canadiens. Although the playoff results aren't considered by voters, the Blackhawks cruised to the Stanley Cup, losing just seven more games along the way.

However, Chicago was seen as a juggernaut that could have won with anybody behind the bench (which is complete nonsense), while Ottawa was viewed as a bad team, having lost their two top players to injury, Erik Karlsson and Jason Spezza, who had made the playoffs anyway, ostensibly because of MacLean's brilliant coaching.

That leads to what has been dubbed the Jack Adams curse. Because the award is usually given to the league's luckiest coach, he tends to lose his job the following season, when that puck luck runs out. That's exactly what happened to MacLean the next season, and the Jack Adams winners for the two following seasons, Colorado's Patrick Roy and Calgary's Bob Hartley.

It's also strange to award the Jack Adams to just the head coach and not the entire staff. That might have made sense in 1973–74, when most NHL coaches stood alone, but now teams have an associate coach, two or three assistant coaches, including a special teams coach and an eye in the sky, a goaltending coach, a special advisor coach, a video coach, a skating coach, a conditioning coach, a development coach, and a leather coach. Okay, maybe not that last one, but you get the idea. These days, coaches can't be effective without the support of their entire staff.

All that brings us to the cold, objective world of hockey analytics, and if there is a single award that is just itching for its assistance, then its the Jack Adams.

Historically, coaches have been judged by their win-loss record. While that isn't a bad approach, it does tend to favour those who were fortunate enough to coach some very good teams. Unless you think that Tom Johnson is history's greatest coach, we should do some fine-tuning. (Tom Johnson, who coached the powerhouse Boston Bruins from 1970–71 to 1972–73, has the best career winning percentage of any NHL coach, 0.738.)

Rather than evaluate a coaching staff based on their end result, we can evaluate them based on the difference between those results and the team's expectations, just like we do for players. Yes, there are other factors that can explain the difference between a team's results and its expectations besides the coaching staff, like a surprise individual breakout (or fall away) season for which the coaches weren't responsible, a particularly hot or cold goalie, or simple puck luck. However, such factors should start to cancel each other out over time, leaving only the portion for which the coaches were responsible.

The only trick is setting those expectations by somehow determining how well each team would have performed with some other coaching staff instead. For example, just how many points would the 2012–13 Blackhawks have earned with a staff other than Quenneville and company?

Theoretically, a team's opening day expectations could be established by going through the various predictions in the popular books, magazines, and newspapers. Not only is that a lot of work, but it's a highly subjective exercise, conducted by pundits without the same access to information as the coaching staff. In essence, we might just be trading one set of biases for another. Besides, expert predictions really aren't that accurate. I tracked them very carefully from 2013–14 through 2015–16, and the results weren't very promising. On the average, the previous season's standings were just as reliable a predictor of the following season's standings as the average expert projection.[87]

That's why my preferred alternative to setting expectations, which I first developed in late 2009 for *Hockey Prospectus*, sets a

---

87. Rob Vollman, "Predicting the Standings," *Hockey Abstract 2016 Update* (self-pub., 2016), PDF, 92–100.

team's opening day expectations using the previous season's data regressed to the average by 35% (as calculated by *Behind the Net*'s Gabriel Desjardins, in a separate 2009 study).[88]

The big advantage of setting expectations using a regression is that it avoids favouring coaches of particularly good or bad teams. In essence, the coach of a great team would have to keep them great and avoid slipping back into the pack, while the coach of a bad team would have to improve the team by more than they would have otherwise improved. Ultimately, that approach should largely neutralize team effects in the evaluation of coaches.

One of the more obvious problems with this approach is that a team's roster can be completely overhauled over the course of a single summer. For example, it doesn't seem fair to set expectations for the 1992–93 Quebec Nordiques based on the 1991–92 team, after they added Steve Duchesne, Ron Hextall, Mike Ricci, Andrei Kovalenko, Scott Young, Martin Rucinsky, and Kerry Huffman over the summer. Something tells me that the Eric Lindros trade had a lot more to do with the team's 52-point improvement than coach Pierre Page and his staff.

That's why it may seem preferable to make some kind of adjustment to the previous season's standings, based on trades, free-agent signings, notable rookies, and retirements that took place over the summer. These adjustments would have to be done without after-the-fact information, since the coach in question may have been at least partially responsible for many of the key trades, injuries, or breakout seasons that occurred.

This would also be a lot of work and may not yield more accurate results. The impact of certain changes tend to be exaggerated, like when the return of a healthy Karlsson and Spezza to Ottawa's lineup in 2013–14 falsely led many pundits (myself included) to predict great things for the Senators that year. In fact, Tom Awad abandoned the

---

88. Rob Vollman, "Howe and Why: Coaches," *Hockey Prospectus* (blog), October 26, 2009, http://www.hockeyprospectus.com/puck/article.php?articleid=328; Gabriel Desjardins, "Last Post on Regression to the Mean, I swear," *Arctic Ice Hockey* (blog), October 21, 2009, http://www.arcticicehockey.com/2009/10/21/1090725/last-post-on-regression-to-the.

inclusion of lineup changes in his team projection model altogether because it didn't increase accuracy.[89]

It would appear that the simplest route is also the best. While there are indeed specific individual circumstances when applying a straight 35% regression is the wrong thing to do, it's reasonable on the whole. For the outliers like the old Nordiques, we can always drill down deeper by hand.

As an example of just how credible this approach may be, I applied this system to rank all 30 active coaches weeks into the 2013–14 season, and it's amazing how accurately it identified the league's least effective coaches.[90] The average rank of the 11 coaches who were fired and never subsequently rehired as a head coach was 23rd and included 10 of the bottom 15 coaches.

The ranking wasn't perfect, however: Hartley was ranked number 11, while Peter DeBoer and Jon Cooper were ranked 24th and 25th, respectively. At first, those three rankings didn't appear to be mistakes, since Hartley won the Jack Adams in 2014–15 and DeBoer was fired midway through the season. However, Hartley was fired at the end of the 2015–16 season and has yet to be rehired in the NHL. That same season, DeBoer was quickly hired by San Jose and took them to the Stanley Cup Final. As for Cooper, he is one of four coaches who has remained with his team this entire time, and he took Tampa Bay to the Stanley Cup Final in 2014–15 and to the Eastern Conference Final in 2015–16. Clearly, he shouldn't have been ranked 25th.

Three glaring mistakes out of 30 really isn't that bad, but it's hard to judge any system on individual cases. Instead, grouping the data together can prove to be quite insightful. For example, in the original *Hockey Abstract* it was discovered that[91]

- coaches who never played in the NHL had significantly more success than coaches who had;

89. Tom Awad, "How Good Is Your Team, Really?," *Hockey Abstract 2017*, 8–29.

90. Rob Vollman, "Ranking the NHL's Best Coaches," Bleacher Report, October 29, 2013, http://bleacherreport.com/articles/1826939-ranking-the-nhls-best-coaches.

91. Rob Vollman, "Who Is the Best Coach?," *Hockey Abstract* (2013), 53–61.

- among coaches who had previously played in the NHL, it doesn't appear to matter how good they were, but those who were defencemen had more coaching success than forwards and goaltenders;
- while over 90% of NHL games have been coached by Canadians, they have no demonstrable statistical superiority over Americans.

More specific to the task at hand, grouping together each tier of individual coaches can help determine if this coaching metric is actually useful in predicting which coaches are more likely to get fired in any given season and even which ones are more likely to make the conference finals.

The following chart groups all 240 coach-seasons over the past eight seasons into six groups of 40 coaches apiece, based on where the metric had them ranked when the season began. Coaches ranked in the bottom half were 37% more likely to be terminated either during or immediately after that season, and teams with a coach ranked in the top half were more than two and a half times more likely to make the conference finals.

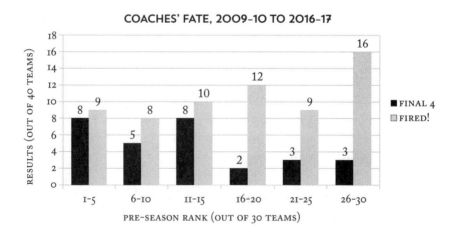

COACHES' FATE, 2009-10 TO 2016-17

Based on this, it seems like this metric does a reasonable job of sorting out which coaches are more likely to do well in the standings and which ones will be fired. But could we achieve these same results just by looking at wins and losses? Do we really need that 35% regression adjustment?

To answer that, I went all the way back to the 2000–01 season, which was the dawn of the 30-team NHL, and I looked for the statistical correlation between where coaches ranked by this metric and how their teams did in the standings. The result of 0.31 is modest but still meaningful, and it is an improvement over games coached (0.18) and has a slight edge over points percentage (0.29). It had a similar but far less significant edge in predicting which coaches would get fired.

Granted, there is a fair bit of selection bias at play, since true Stanley Cup contenders rarely gamble on coaches who don't already have a successful track record. Likewise, weaker teams find it difficult to attract good coaches and are more likely to take a chance on a less experienced coach. Maybe that's why the struggling Edmonton Oilers forged ahead with coaches like Tom Renney, Ralph Krueger, Dallas Eakins, and Todd Nelson before suddenly enticing an established NHL coach like Todd McLellan to town as soon as they won the Connor McDavid lottery.

As for year-over-year persistence, a coach's previous career results aren't terribly predictive of what the coach's results will be for any given season. And so, while this new metric is far from flawless, it is decent enough to be considered a solid first step into the statistical evaluation of coaches.

How Valuable Are Coaches?

At the top of the chapter, I sold you on this study of coaches not just on the premise that our subjective ability to recognize good coaches could use some help, but that coaches can have a meaningful impact on the standings. I've probably convinced you of the former, but how about the latter? Most fans certainly believe it. Based on a Twitter poll of 500 fans that I ran at the time of writing, 57% of fans believed that an elite coaching staff can consistently add 6 to 10 points in the standings, 22% believed that it can add 11 to 15 points, and 9%

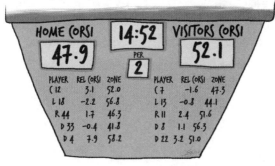

―ADVANCED STATS SCOREBOARD―

HOME CORSI	14:52	VISITORS CORSI
47.9	PER **2**	**52.1**

PLAYER	REL CORSI	ZONE		PLAYER	REL CORSI	ZONE
C 12	3.1	52.0		C 7	-1.6	47.3
L 18	-2.2	56.8		L 13	-0.8	44.1
R 44	1.7	46.3		R 11	2.4	51.6
D 33	-0.4	41.8		D 8	1.1	56.3
D 4	7.9	58.2		D 22	3.2	51.0

"SO, UH... ARE WE WINNING?"

believed it can add even more. Only 12% of fans felt that their impact was 5 points or less.[92]

If that sentiment is accurate, then a coaching staff that can consistently add nine points to their team's rank in the standings every season would be worth around $13.5 million dollars, as per the method of converting dollars to points that we used on players in the opening chapter of *Stat Shot*.[93] That would make an elite coach the most valuable player on the team, save for perhaps the goalie or a marquee player like Connor McDavid. However, there are few teams who pay their coaching staff that much, if any.

What about the opinion of the NHL's front offices? Their points of view can be captured by reverse engineering the investments they make in coaching. While salaries of assistant coaches aren't generally known, the average salary of a head coach going into the 2016–17 season was $3.23 million, based on the 14 coaches whose salaries are known.[94] That's almost exactly the same as the average salary for a player; given the salary cap of $73 million and a roster of 23 players, you get $3.17 million.

If NHL GMs are right, then the head coach is only worth a couple of points in the standings, and presumably the rest of the

92. Rob Vollman (@robvolllmanNHL), "On average, how many points per season do you think an elite NHL coaching staff can add to the standings, over the long run," Twitter, May 30, 2017, 7:21 a.m., https://twitter.com/robvollmanNHL/status/869559335285694464.

93. Rob Vollman, "What's the Best Way to Build a Team?," *Stat Shot*, 36–40.

94. Coaches' salaries data from Cap Friendly, accessed June 4, 2017, https://www.capfriendly.com/coaches.

staff is worth even less. As such, there's a decent gap in the perceived value of an elite coaching staff between the conventional wisdom of fans and the practical applications in front offices.

So who is right? Based on the annual results since I first introduced this metric back in 2009, the truth may fall roughly halfway in between.[95] A good coaching staff can add between three and six points to their team's rank in the standings on a regular basis, and it's not unusual for the best of them to add up to eight or ten points in seasons when everything comes together.

To take some recent examples, the following leaderboard of everyone who has served as an NHL coach at some point in the past five seasons highlights how Bruce Boudreau's teams have earned 1,004 points in 763 games coached (GC), which is 91.5 points above expectations (PAX), which works out to 9.8 points per 82-game season (PAX/82). To place his career totals in historical context, that ranks fifth behind Bowman (291.6), Arbour (109.0), Glen Sather (102.0), Dick Irvin (96.7), and Dallas coach Ken Hitchcock (93.1). It's no wonder Minnesota improved from 87 to 106 points in 2016–17.

ALL-TIME RESULTS FOR RECENTLY ACTIVE NHL COACHES[96]

TEAM	COACH	GC	PTS	PAX	PAX/82
Minnesota	Bruce Boudreau	763	1,004	91.5	9.8
New York Islanders	Doug Weight	40	52	4.7	9.6
inactive	Dan Bylsma	565	695	46.4	6.7
Tampa Bay	Jon Cooper	344	413	24.7	5.9
Dallas	Ken Hitchcock	1,454	1,761	93.1	5.2
Edmonton	Todd McLellan	704	861	40.3	4.7
Chicago	Joel Quenneville	1,539	1,903	86.4	4.6
inactive	Patrick Roy	246	284	13.3	4.4
Montreal	Claude Julien	1,021	1,236	55.4	4.4
Vegas	Gerard Gallant	327	339	17.0	4.3
Nashville	Peter Laviolette	1,005	1,160	51.0	4.2

95. Rob Vollman, "Howe and Why: Coaches," *Hockey Prospectus* (blog), October 26, 2009, http://www.hockeyprospectus.com/puck/article.php?articleid=328.

96. Coaching raw data for calculations from Hockey Reference, accessed June 4, 2017, http://www.hockey-reference.com.

TEAM	COACH	GC	PTS	PAX	PAX/82
Toronto	Mike Babcock	1,114	1,356	56.8	4.2
inactive	Michel Therrien	814	926	37.6	3.8
Washington	Barry Trotz	1,442	1,613	66.3	3.8
NY Rangers	Alain Vigneault	1,134	1,352	47.6	3.4
inactive	Darryl Sutter	1,285	1,435	49.6	3.2
Ottawa	Guy Boucher	278	312	7.9	2.3
inactive	Paul MacLean	239	263	5.5	1.9
Pittsburgh	Mike Sullivan	300	360	6.9	1.9
inactive	Bob Hartley	944	1,046	21.4	1.9
Anaheim	Randy Carlyle	786	913	16.9	1.8
Philadelphia	Dave Hakstol	164	184	2.9	1.5
St. Louis	Mike Yeo	381	436	4.9	1.1
Calgary	Glen Gulutzan	212	231	2.5	1.0
Vegas (AHL)	Craig Berube	161	178	1.6	0.8
Columbus	John Tortorella	1,093	1,191	8.4	0.6
Calgary-Asst	Dave Cameron	137	157	0.8	0.5
NY Rangers-Asst	Lindy Ruff	1,493	1,675	4.8	0.3
Boston	Bruce Cassidy	137	147	0.3	0.2
inactive	Jack Capuano	483	518	−0.2	0.0
inactive	Dave Tippett	1,114	1,254	−4.1	−0.3
San Jose	Peter DeBoer	658	708	−6.2	−0.8
inactive	Adam Oates	130	147	−2.1	−1.3
Tampa Bay-Asst	Todd Richards	424	445	−10.6	−2.0
Montreal-Asst	Kirk Muller	187	187	−5.0	−2.2
Winnipeg	Paul Maurice	1,365	1,392	−38.7	−2.3
Boston-Asst	Joe Sacco	294	290	−9.1	−2.5
Carolina	Bill Peters	246	244	−8.1	−2.7
inactive	Ralph Krueger	48	45	−1.9	−3.3
WHL	Mike Johnston	110	131	−4.6	−3.4
New Jersey (org.)	Claude Noel	201	204	−8.4	−3.4
Chicago-Asst	Kevin Dineen	146	140	−7.4	−4.2
Detroit-Asst	John Torchetti	66	66	−3.7	−4.6
inactive	Ted Nolan	471	431	−34.3	−6.0
Detroit (AHL)	Todd Nelson	51	43	−4.0	−6.4
inactive	Willie Desjardins	246	245	−19.4	−6.5

TEAM	COACH	GC	PTS	PAX	PAX/82
New Jersey	John Hynes	164	154	−15.4	−7.7
Detroit	Jeff Blashill	164	172	−17.5	−8.8
inactive	Ron Rolston	51	44	−10.8	−17.3
New Jersey (org.)	Peter Horachek	108	79	−23.3	−17.7
Anaheim (AHL)	Dallas Eakins	113	86	−24.6	−17.8
Florida (org.)	Tom Rowe	61	58	−15.6	−21.0
Colorado	Jared Bednar	82	48	−37.3	−37.3

As is the custom in these pages, it's time for a quick sanity check: the results of an objective analysis should be compared to a subjective selection of the league's best coaches.

Mike Johnston of Sportsnet ranked the league's 30 coaches prior to the 2015–16 season, and he included Mike Babcock, Quenneville, Barry Trotz, Alain Vigneault, Hartley, and Boudreau in the elite tier.[97] A year later, Sean O'Leary of the Score ranked the coaches, and he included Quenneville, Trotz, Cooper, Darryl Sutter, Babcock, and Lindy Ruff in his top six.[98]

Of the three coaches who appeared in Johnston's elite tier and O'Leary's top six, Quenneville ranks sixth in PAX/82, Babcock ranks 10th, and Trotz is 11th among today's 30 active coaches. Actually, make that 31 active coaches, since the league now includes the Vegas Golden Knights, who named Gerard Gallant, who ranks eighth, as their new coach.

I'd prefer a stronger sanity check than 6th, 10th, and 11th, but consider the coaches above them. Boudreau ranks first, Cooper ranks third, and they were included at the top of Johnston and O'Leary's lists, respectively. Doug Weight ranks second, but he has only coached 40 games, so I skipped him in the rankings. A more recent subjective list would probably include McLellan and Peter Laviolette near the top, given how Edmonton went from a laughing

97. Mike Johnston, "NHL Coaching Power Rankings: Seeding All 30 Bench Bosses," Sportsnet, October 7, 2015, http://www.sportsnet.ca/hockey/nhl/nhl-coaching-power-rankings-seeding-all-30-bench-bosses/.

98. Sean O'Leary, "Ranking the 30 NHL Head Coaches," The Score, August 2, 2016, https://www .thescore.com/nhl/news/1064618-ranking-the-30-nhl-head-coaches.

stock to a video review away from the Western Conference Final, and how Nashville made a surprise trip to the Stanley Cup Final.

As for Hitchcock and Claude Julien, I have no idea why they weren't in either top six, given their clear and obvious career coaching achievements. In fact, three of the top subjectively ranked coaches aren't NHL head coaches anymore: Hartley, Sutter, and Ruff. I'm sure they'll get jobs by the time this book goes to print, but it demonstrates how the subjectively ranked list isn't a foolproof way of identifying good coaches either. Given that this entire chapter was introduced on the premise that we're not equipped to make particularly accurate subjective assessments of coaches, it shouldn't be terribly surprising nor discouraging that we didn't pass the sanity check with flying colours.

To me, the more pressing issue is the lack of data on which to judge the coaches. There's an urge to run to the bank with some of these results, but it would be nice to judge each coach's performance over a longer record that spans several teams and leagues.

Outside the NHL

Every season, there are several coaches who have been successful in other leagues who are given the reins of an NHL team for the first time. Recently, these included Willie Desjardins, Dave Cameron, Bill Peters, Todd Nelson, Mike Johnston, and Peter Horachek in 2014–15, John Hynes, Jeff Blashill, and Dave Hakstol in 2015–16, Tom Rowe and Jared Bednar in 2016–17, and Travis Green, Bob Boughner, and Phil Housley in 2017–18. Accurately evaluating a team's coaching requires knowing how well such coaches performed outside the NHL.

Thanks to Josh Weissbock (now of the Florida Panthers) gathering most of the historical data a few years ago, I have been able to incorporate coaching data from the AHL, the three Canadian major junior leagues, U.S. college hockey, and even the ECHL into this model. In the future, it would be great to add European leagues, like the KHL and Switzerland's National League A.

The data quality isn't perfect and it took a lot of manual effort to clean it up, especially for the NCAA and ECHL, but it has opened the door to learning more about the history behind today's NHL coaches, and who may potentially join them in the future.

If we apply this new perspective to the current NHL coaches, the extra data helps paint a slightly more confident picture. In the following table, I very arbitrarily decided to divide each coach's non-NHL PAX by half for the final PAX/82 result, given that success in the NHL is at least twice as impressive as equivalent success somewhere else.

CURRENT NHL COACHES, 2017–18[99]

COACH	TEAM	NHL GC	NHL PAX	OTHER GC	OTHER PAX	PAX/82
Doug Weight	NYI	40	4.7	0	0.0	9.6
Bruce Boudreau	MIN	763	91.5	865	75.4	7.6
Jon Cooper	TBL	344	24.7	221	47.4	6.6
Gerard Gallant	VGK	327	17.0	204	58.3	6.0
Ken Hitchcock	DAL	1,454	93.1	432	65.6	5.4
Joel Quenneville	CHI	1,539	86.4	80	1.4	4.5
Peter Laviolette	NSH	1,005	51.0	230	32.0	4.3
Todd McLellan	EDM	704	40.3	752	52.5	4.1
Claude Julien	MTL	1,021	55.4	487	23.6	4.0
Mike Babcock	TOR	1,114	56.8	736	35.2	3.6
Barry Trotz	WSH	1,442	66.3	400	23.8	3.6
Mike Sullivan	PIT	300	6.9	94	29.7	3.4
Dave Hakstol	PHI	164	2.9	459	43.2	2.9
Travis Green	VAN	0	0.0	350	23.2	2.7
Alain Vigneault	NYR	1,134	47.6	741	4.2	2.7
Guy Boucher	OTT	278	7.9	430	27.3	2.4
Glen Gulutzan	CGY	212	2.5	593	44.3	2.2
Randy Carlyle	ANA	786	16.9	80	13.7	2.0
Peter DeBoer	SJS	658	−6.2	878	117.0	1.7
Mike Yeo	STL	381	4.9	80	14.9	1.7

99. NHL coaching raw data for calculations from Hockey Reference, accessed June 4, 2017, http://www.hockey-reference.com; non-NHL coaching raw data from Elite Prospects, accessed June 4, 2017, http://www.eliteprospects.com, gathered by Josh Weissbock.

COACH	TEAM	NHL GC	NHL PAX	OTHER GC	OTHER PAX	PAX/82
Bruce Cassidy	BOS	137	0.3	730	24.5	1.0
John Tortorella	CBJ	1,093	8.4	160	8.3	0.7
Bob Boughner	FLA	0	0.0	544	9.3	0.7
Phil Housley	BUF	0	0.0	0	0.0	0.0
Bill Peters	CAR	246	−8.1	456	24.9	−0.3
John Stevens	LAK	267	−6.4	480	−9.5	−1.4
John Hynes	NJD	164	−15.4	384	33.7	−1.6
Paul Maurice	WPG	1,365	−38.7	212	21.2	−1.9
Jeff Blashill	DET	164	−17.5	228	28.2	−3.1
Rick Tocchet	ARI	148	−5.9	0	0.0	−3.3
Jared Bednar	COL	82	−37.3	451	21.4	−8.5

In addition to some minor shuffling of some of the more established coaches, the biggest change was being able to rank coaches without NHL experience, like Green and Boughner, and being able to moderate the results of those with very little, like Bednar, Blashill, Hynes, Cassidy, and Hakstol. Nothing can be done about Weight or Housley, who lack experience at any level.

However, we must concede that a coach's job outside the NHL isn't necessarily to win but to develop the team's players. Outside the NCAA, coaches also have very little control over their rosters, which are subject to frequent change that can quickly sabotage any attempt to accurately establish the team's expectations. However, the primary duty of an NHL coach is to win, and hopefully this outside data is useful in establishing which coaches have done exactly that, regardless of the circumstances.

We also have yet to determine if there is a useful connection between coaching success in other leagues and the NHL. Clearly, there will be some selection bias at play, since only the coaches who were successful in other leagues will be offered an NHL head coaching assignment. However, if the connection is there, then we can use this data to more accurately evaluate a coach's experience and prior success. If the connection isn't there, then this data is

nothing more than interesting trivia, and we'll have to limit ourselves to NHL data.

To evaluate the connection, I compiled the following chart,
which places each coach's career AHL PAX/82 on the horizontal
axis and his NHL results on the vertical axis, much as we have previously done with translation factors. (Minimum 100 games coached
in both the AHL and the NHL.) If there's a connection between
the two, then the points will roughly cluster along a line. In my view,
there is a slight relationship between the two sets of data, but not
one that is strong enough to be clearly graphed.

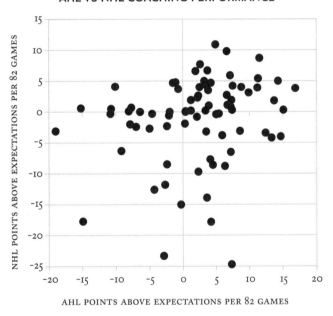

AHL VS NHL COACHING PERFORMANCE

AHL POINTS ABOVE EXPECTATIONS PER 82 GAMES

Focusing on what's to the left of the vertical line, the 25 coaches
who were below zero in the AHL were an average of 2.5 points below
expectations in the NHL, while the 52 to the right of the line more
or less broke even in the NHL. Focusing on the far right, the 15 best
AHL coaches added an average of 1.7 points per 82-game season
in the NHL. Admittedly, the relationship between AHL coaching
success and success in the NHL is quite modest, but it does exist, so
we can use this data.

Beyond confirming that we can include this data in our evaluation of active NHL coaches, the other advantage of establishing the value of AHL data is that it can help identify the best future coaching options. Let's dig a little deeper and see who we can find.

American Hockey League

When Mike Sullivan was named coach of the Pittsburgh Penguins on December 12, 2015, they were in the middle of the standings with a 15-10-3 record and had just come off their worst season since 2005–06, when Sidney Crosby was a rookie. Since then, they have the second-best record in the NHL and won the Stanley Cup both seasons. Maybe it's a coincidence, or maybe Sullivan deserves a lot of credit for so quickly and completely reversing the team's fortunes.

Those who believe the latter explanation won't be surprised to see Sullivan at the top of the following leaderboard, which includes the career AHL data for the top 30 coaches who have served as a head coach in either the NHL or the AHL in the past five seasons. While the sample size is admittedly small, Sullivan's success with the Providence Bruins in 2002–03 and the Scranton Wilkes-Barre Penguins in 2015–16 certainly foreshadowed his success in Pittsburgh.

TOP-30 ACTIVE COACHES, ALL-TIME AHL RESULTS[100]

TEAM	COACH	GC	PTS	PAX	PAX/82
Pittsburgh	Mike Sullivan	94	131	29.7	25.9
Vegas (AHL)	Craig Berube	161	205	45.8	23.3
inactive	Willie Desjardins	152	203	37.3	20.1
Tampa Bay	Jon Cooper	221	292	47.4	17.6
Columbus-Asst	Brad Larsen	152	199	24.9	13.4
Minnesota (AHL)	Derek Lalonde	75	79	12.0	13.1
Detroit	Jeff Blashill	228	291	28.2	10.2
NY Islanders (AHL)	Brent Thompson	303	334	33.4	9.0
NCHC	Mike Haviland	316	388	33.3	8.6

100. Coaching raw data for calculations from Elite Prospects, accessed June 4, 2017, http://www .eliteprospects.com, gathered by Josh Weissbock.

TEAM	COACH	GC	PTS	PAX	PAX/82
Tampa Bay (AHL)	Benoit Groulx	235	245	22.4	7.8
Toronto (AHL)	Sheldon Keefe	151	203	13.4	7.3
Anaheim (AHL)	Dallas Eakins	447	528	39.2	7.2
New Jersey	John Hynes	384	493	33.7	7.2
Washington (AHL)	Troy Mann	227	293	19.7	7.1
Florida (org.)	Tom Rowe	426	479	36.3	7.0
Boston (AHL)	Kevin Dean	75	96	5.9	6.5
Boston	Bruce Cassidy	460	554	35.6	6.3
Minnesota (AHL) Asst	David Cunniff	26	23	1.8	5.6
Philadelphia (AHL)	Scott Gordon	560	663	38.2	5.6
Carolina (AHL)	Ulf Samuelsson	75	85	4.5	5.0
San Jose-Asst	Steve Spott	76	96	4.1	4.4
inactive	Troy Ward	228	264	11.8	4.2
Pittsburgh (AHL)	Clark Donatelli	124	154	6.2	4.1
Minnesota-Asst	John Anderson	788	942	37.2	3.9
ECAC	Mark Morris	704	822	32.3	3.8
Nashville (AHL)	Dean Evason	379	455	16.8	3.6
Detroit (AHL)	Todd Nelson	484	586	20.1	3.4
Buffalo-Asst	Terry Murray	353	329	10.8	2.5
Vancouver	Travis Green	303	349	8.4	2.3
Colorado	Jared Bednar	308	357	4.4	1.2

Before getting too excited about Sullivan, there were three other coaches who enjoyed comparable success in the AHL but without success in the NHL: Craig Berube, Willie Desjardins, and Jeff Blashill. In addition, some of those with more modest AHL success, like Dallas Eakins, John Hynes, and Tom Rowe, didn't exactly set the league on fire when they had their brief opportunities with Edmonton, New Jersey, and Florida, respectively. Sullivan's only fellow NHL success story is Cooper. That doesn't bode well for the Boston Bruins, Vancouver Canucks, and Colorado Avalanche, whose coaches (Bruce Cassidy, Travis Green, and Jared Bednar) are all even farther down the list.

Other than the AHL, where 61 of last year's head and assistant coaches had previously plied their trade, the three Canadian major junior leagues are the primary source of NHL coaching talent, with 45.

Of the NHL's 30 head coaches, 13 had previously coached in the CHL. Among the three Canadian major junior leagues, OHL coaches have always been a stride behind those of the WHL and QMJHL. For whatever reason, OHL coaching success hasn't translated to the NHL, whether it's historical legends like Dick Todd, Brian Kilrea, and Bert Templeton or more recent stars like Dale Hunter and Dave Cameron. Until Peter DeBoer and his OHL-intensive staff overcame their initial struggles in Florida and New Jersey and finally broke out for the San Jose Sharks in 2015–16, NHL success for an OHL coach was basically without precedent.

The same cannot be argued for coaches from the WHL or QMJHL; several have made the transition, including Mike Babcock, Ken Hitchcock, and Todd McLellan in the former case, and Alain Vigneault, Guy Boucher, Claude Julien, and Michel Therrien in the latter.

Given how many great coaches have come from the CHL, it makes sense to study the career numbers of the top coaches who have been active in at least one of these leagues over the past five seasons to see who might be next.

TOP-50 ACTIVE CHL COACHES, ALL TIME[101]

LEAGUE(S)	COACH	GC	PTS	PAX	PAX/82
WHL	Emanuel Viveiros	72	88	22.9	26.1
WHL	Travis Green	47	76	14.8	25.9
QMJHL	Ross Yates	68	74	20.2	24.3
WHL	Brent Kisio	144	189	41.9	23.8
OHL	Kris Knoblauch	457	625	84.3	22.7
WHL	Malcolm Cameron	72	85	19.2	21.9

101. Coaching raw data for calculations from Elite Prospects, accessed June 4, 2017, http://www.eliteprospects.com, and gathered by Josh Weissbock.

LEAGUE(S)	COACH	GC	PTS	PAX	PAX/82
WHL	Richard Matvichuk	72	96	18.5	21.1
WHL	Derek Laxdal	288	387	63.5	18.1
OHL	Sheldon Keefe	243	319	51.5	17.4
OHL	Stan Butler	1,496	1575	69.2	17.4
OHL	D.J. Smith	262	353	55.1	17.3
WHL	Mike Johnston	443	573	89.3	16.5
WHL	Kevin Constantine	576	720	111.0	15.8
OHL	Jacques Beaulieu	344	354	28.3	13.8
QMJHL	Louis Robitaille	68	78	11.2	13.5
WHL	Dave Lowry	432	567	65.1	12.4
WHL	Ryan Huska	504	643	75.4	12.3
WHL	John Paddock	216	277	32.2	12.2
QMJHL	Martin Bernard	381	395	54.8	11.8
WHL	Tim Hunter	216	243	30.1	11.4
OHL, QMJHL	Jim Hulton	657	656	8.3	11.3
OHL	Bill Stewart	161	187	21.4	10.9
WHL	Lorne Molleken	1,078	1,290	143.7	10.9
OHL	Jeff Brown	204	216	24.1	9.7
OHL	Dale Hunter	1,028	1,453	118.1	9.4
QMJHL	Éric Veilleux	600	739	68.6	9.4
QMJHL	J.F. Houle	305	378	34.5	9.3
WHL	Dan Lambert	72	112	7.9	9.0
OHL	Jay McKee	68	77	7.4	8.9
QMJHL	Patrick Roy	545	735	53.7	8.1
QMJHL	Gilles Bouchard	272	352	25.5	7.7
OHL	Drew Bannister	136	174	12.6	7.6
QMJHL	Dominic Ricard	205	212	18.5	7.4
OHL	Marty Williamson	816	977	66.9	6.7
WHL	Don Hay	1,296	1,558	104.8	6.6
WHL	Marc Habscheid	842	931	68.0	6.6
WHL, OHL	Ryan McGill	712	802	39.0	6.5
QMJHL	Benoit Groulx	902	1,072	65.5	6.0
OHL	Scott Walker	306	380	21.7	5.8
WHL	Guy Charron	428	457	9.2	5.7
QMJHL	Mario Durocher	803	956	53.0	5.4

LEAGUE(S)	COACH	GC	PTS	PAX	PAX/82
WHL	Steve Young	271	274	17.6	5.3
QMJHL	Dominique Ducharme	408	484	21.2	4.8
QMJHL	Yanick Jean	771	855	44.4	4.7
WHL	Jesse Wallin	311	339	17.5	4.6
WHL	Steve Konowalchuk	432	475	23.8	4.5
WHL	Cory Clouston	576	683	30.6	4.4
WHL	Brent Sutter	913	1,087	40.9	3.7
WHL	Don Nachbaur	1,338	1,523	59.0	3.6
WHL	Dean Clark	987	1,036	42.7	3.5

Of the seven coaches who added more than 20 points in an 82-game schedule, only Kris Knoblauch has a sufficient sample size, having served as a WHL assistant coach for three seasons and then as the head coach of the Kootenay Ice for two seasons, including the championship 2010–11 season, and most recently the Erie Otters for five seasons, including the 2016–17 championship.

Only 38 years old, Knoblauch has helped his team win at least 50 of the league's 68 scheduled games in each of the past four seasons and compile an 11-3 series record. Yes it is the OHL and yes he has had stars like Connor McDavid, Dylan Strome, and Alex DeBrincat in his lineup, but those are still awfully impressive results. That's probably why he was signed as an assistant coach by the Philadelphia Flyers in the summer of 2017.

Another young name is former Tampa Bay Lightning depth forward Sheldon Keefe, who is 36 years old. In two seasons apiece with the Sault Ste. Marie Greyhounds of the OHL and the Toronto Marlies of the AHL, Keefe has compiled a jaw-dropping record of 194-74-20.

In terms of more established and legendary names there's Stan Butler, who has coached the North Bay and Brampton Battalion for the past 19 seasons. They may not have won any championships and their 2016–17 season was certainly forgettable, but the team has otherwise always been competitive.

Getting away from the OHL, some of the top WHL names on

this list with decent sample sizes are already recognizable within the NHL. Mike Johnston was an assistant with Vancouver and Los Angeles before coaching the Penguins; Kevin Constantine coached the Sharks, Penguins, and Devils long ago; Dave Lowry was named an assistant coach for the Kings after a previous three-year stint with the Flames; John Paddock coached the first iteration of the Winnipeg Jets as well as the Senators and was an assistant coach for the Flyers; and WHL legend Lorne Molleken of the Saskatoon Blades was briefly Chicago's head coach in the late 90s, as well as a short-term assistant with San Jose and Pittsburgh a few years later.

As for some of the lesser-known names, Derek Laxdal has 15 years of experience as a head coach in the Central Hockey League, the ECHL, the WHL, and the AHL, but his greatest success was in the WHL. The Edmonton Oil Kings won at least 50 of 72 games in the final three of the four seasons he coached them, and they compiled an 11-1 series record in the playoffs. Since then, Laxdal has had a record of 114-84-30 with the AHL's Texas Stars.

And then there's Ryan Huska, who's 41 and was with the Kelowna Rockets for 12 seasons, the last seven as the head coach. He won the championship in 2008–09 and had an amazing 109-27-8 record over his final two seasons with the team. The Calgary Flames scooped him up, and he's been the head coach of the Adirondack Flames and Stockton Heat since then, albeit with mixed results.

As for the QMJHL, it's surprising to see Martin Bernard's name so high on the list, with just 395 points in his 381 games coached and having never won a playoff series. However, he's the perfect example of why this coaching metric is preferable over just looking at unadjusted wins and losses: in 2006–07, the Victoriaville Tigres improved from 26-42-2 to 38-25-7, in 2013–14 the Shawinigan Cataractes improved from 15-46-7 to 20-39-9, and in 2016–17 the Baie-Comeau Drakkar improved from 14-49-5 to 26-32-10. It may not be clear what some of those team names mean, but it is clear that Bernard has dramatically improved his teams in all three coaching opportunities.

Elsewhere in the QMJHL, several coaches have stats similar to Patrick Roy—who is known to have been a reasonably effective NHL coach with the Colorado Avalanche—including Gatineau

Olympiques legend and AHL Syracuse Crunch head coach Benoit Groulx, Éric Veilleux, and Jean-François Houle, with an eye on Gilles Bouchard and Dominic Ricard.

U.S. College Hockey

There are some challenges that are unique to evaluating coaches in U.S. college hockey. On the one hand, their rosters are relatively static, since they're composed not of single-focused hockey players but of students who generally want to get an education and stay in the same program for all four or five seasons without changing teams. On the other hand, that also means that the emphasis is a little bit less on winning and more on development and scholastic achievement.

There's also the added issue of data quality, as various data sources will be in direct conflict about who coached a particular team at any given time, and it has to be sorted out by hand. For example, Hockey DB correctly has Nathan Leaman listed as the coach of Union College from 2003–04 to 2010–11, while Elite Prospects previously had it listed as Tim Gerrish and/or Rick Bennett (this has since been corrected).[102]

There's also unbalanced schedules, varying quality of conferences, a short schedule, and a whole host of other obstacles that make it difficult to evaluate NCAA coaches. Of them all, the greatest issue could be the lack of precedent against which to compare today's coaches. Last season, there were only six individuals on NHL coaching staffs that had previously coached in NCAA Division I, and in all but one case it was a very brief experience and/or as an assistant coach. In fact, two of those six coaches were not renewed for the 2017–18 season.

The one prominent exception is the former 15-year veteran coach of the University of North Dakota Fighting Sioux, David Hakstol. On May 18, 2015, he was given the head coaching reins of the Philadelphia Flyers (perhaps only because his last name sounds

102. HockeyDB, accessed June 4, 2017, http://www.hockeydb.com/ihdb/stats/pdisplay.php?pid=56437; Elite Prospects, accessed June 4, 2017, http://www.eliteprospects.com/team.php?team=1366&year0=2006.

like GM Ron Hextall's). Since then, I think it's fair to say that he has done reasonably well with the pretty mediocre lineup he was given.

My proposed solution to resolve this uncertainty is to use the two most legendary coaches in league history as a basis of comparison, each of whom recently retired, Jack Parker and Red Berenson. After 39 seasons as the head coach of the Boston University Terriers, which was preceded by five seasons as a player and/or assistant coach, Jack Parker retired at the end of the 2012–13 season. Then there's NHL legend Red Berenson, who won the Jack Adams with the Blues in 1980–81 and then served as an assistant for the Sabres under Bowman. He retired in 2015–16 after coaching the University of Michigan Wolverines for 32 seasons. Put together, Berenson and Parker can serve as a frame of reference against which to judge today's top coaches.

Using the following table, let's define the Berenson-Parker line as being somewhere between the 8.9 and 10 extra points they gave their teams over an 82-game schedule. Yes, NCAA schedules are far shorter than that, but using 82 games allows us to compare across leagues. NCAA Division I is a real untapped market for coaches, but anyone who is above the Berenson-Parker line could be a leading candidate as Hakstol helps open minds across the NHL.

TOP-30 ACTIVE U.S. COLLEGE COACHES, ALL-TIME[103]

COACH	GC	PTS	PAX	PAX/82
Tony Granato	36	41	12.4	28.1
Jim Montgomery	124	178	32.2	21.3
David Quinn	119	159	30.8	21.2
Norm Bazin	281	378	68.5	20.0
Mike Guentzel	152	207	32.7	17.6
Mike Hastings	202	262	35.4	14.4
Nate Leaman	231	285	29.8	10.6
Brett Larson	37	41	4.6	10.2
Jack Parker	1,413	1,832	172.7	10.0
Jeff Jackson	690	899	84.1	10.0
Rick Bennett	371	465	41.7	9.2

---

103. Coaching raw data for calculations from Elite Prospects, accessed June 4, 2017, http://www .eliteprospects.com, gathered by Josh Weissbock.

COACH	GC	PTS	PAX	PAX/82
Rand Pecknold	683	842	76.0	9.1
Red Berenson	1,350	1,767	146.3	8.9
Don Lucia	1,110	1,365	114.3	8.4
Mike Schafer	639	784	65.1	8.4
Mark Morris	538	695	54.7	8.3
Dean Blais	1,008	1,262	97.4	7.9
David Hakstol	459	597	43.2	7.7
Robert Ferraris	122	136	11.0	7.4
George Gwozdecky	956	1,117	82.6	7.1
Derek Schooley	259	283	20.3	6.4
Wayne Wilson	218	218	16.5	6.2
Mike Cavanaugh	512	634	32.1	5.1
Andy Murray	192	195	10.8	4.6
Enrico Blasi	709	817	39.9	4.6
Richard Umile	734	871	41.1	4.6
Keith Allain	412	479	22.7	4.5
Bob Motzko	507	571	27.4	4.4
Guy Gadowsky	577	553	30.5	4.3
Chris Bergeron	274	265	14.4	4.3

Unsurprisingly, few coaches reach or exceed the Berenson-Parker line with any reasonable sample size, and Hakstol isn't even one of them. However, there are plenty of established NCAA coaches who are right up there and warrant a closer look. Let's hope that some of them get NHL opportunities and provide us with a greater volume of information for future work.

ECHL

If there isn't enough quality data to evaluate U.S. college coaching, then I must really be out of my mind to dive into the ECHL. However, there were 14 coaches on staff in the NHL last season who had previously coached in the ECHL, including head coaches

Jack Capuano, Peter Laviolette, Bruce Boudreau, Jared Bednar, Glen Gulutzan, and Bruce Cassidy. The ECHL is not to be ignored.

What is the ECHL? Just like the AHL, it's a North American professional hockey league with NHL affiliates, but the quality of competition is just a step below. Players and coaches rarely go directly from the ECHL to the NHL, but lots of future NHLers are initially developed there. Formerly known as the East Coast Hockey League, the ECHL goes back to 1988–89. It merged with the West Coast Hockey League in 2003–04 and the Central Hockey League in 2014–15.

Right now, the top ECHL coach is probably Steve Martinson. If you include his time in the WCHL and CHL, then his teams have won 10 championships over his 21 years as a head coach, and he has a record of 198-95-37 with his current team, the Allen Americans.

TOP-20 ACTIVE ECHL COACHES, ALL-TIME[104]

COACH	GC	PTS	PAX	PAX/82
Derek Lalonde	144	206	49.1	27.9
Aaron Schneekloth	72	99	15.3	17.5
Richard Matvichuk	144	174	29.5	16.8
Dan Watson	72	106	13.9	15.8
Steve Martinson	576	744	95.3	13.6
Brad Ralph	288	381	43.7	12.4
John Wroblewski	144	182	21.8	12.4
Colin Chaulk	144	145	20.6	11.7
Gary Graham	288	373	37.7	10.7
Malcolm Cameron	627	749	79.3	10.4
Cail MacLean	288	349	35.6	10.1
Troy Mann	216	222	24.9	9.5
Dean Stork	360	430	34.4	7.8
Spencer Carbery	360	452	31.0	7.1
Jamie Russell	144	150	11.9	6.8
Bruce Ramsay	72	80	5.6	6.3
Larry Courville	614	748	40.4	5.4

104. Coaching raw data for calculations from Elite Prospects, accessed June 4, 2017, http://www .eliteprospects.com, and gathered by Josh Weissbock.

COACH	GC	PTS	PAX	PAX/82
Greg Poss	432	535	27.1	5.1
Terry Ruskowski	412	457	22.7	4.5
Vince Williams	288	299	13.7	3.9

Top Coaches Outside the NHL

Add up all of this information from the AHL, CHL, NCAA, and ECHL, and the result is the following table of all the top coaches outside the NHL who have been active over the course of the past five seasons. (Well, some of them are currently serving as assistant NHL coaches or recently served as a head coach.) Remember, I have very arbitrarily chosen to cut each coach's non-NHL PAX in half. In the future, more exact translation factors will need to be established.

TOP-30 HEAD COACH CANDIDATES, ACTIVE IN 2017–18[105]

COACH	NHL GC	NHL PAX	OTHER GC	OTHER PAX	PAX 82
Kris Knoblauch	0	0	457	84.3	7.6
Steve Martinson	0	0	576	95.3	6.8
Sheldon Keefe	0	0	394	64.9	6.8
Dan Bylsma	565	46.4	54	7.7	6.7
Kevin Constantine	377	30.3	816	120.2	6.3
Dave Lowry	0	0	432	65.1	6.2
Martin Bernard	0	0	381	54.8	5.9
Malcolm Cameron	0	0	699	98.5	5.8
Lorne Molleken	47	1.8	1,318	157.0	4.8
Rand Pecknold	0	0	683	76.0	4.6
Craig Berube	161	1.6	161	45.8	4.4
Michel Therrien	814	37.6	717	102.9	4.4
Mike Johnston	110	−4.6	443	89.3	4.4
Terry Murray	1,012	59.6	353	10.8	4.3
Jeff Jackson	0	0	872	91.2	4.3

---

105. Coaching raw data for calculations from Elite Prospects, accessed June 4, 2017, http://www .eliteprospects.com, gathered by Josh Weissbock.

COACH	NHL GC	NHL PAX	OTHER GC	OTHER PAX	PAX 82
Patrick Roy	246	13.3	545	53.7	4.2
Don Lucia	0	0	1,110	114.3	4.2
Mike Schafer	0	0	639	65.1	4.2
Troy Mann	0	0	443	44.6	4.1
Dean Blais	0	0	1,008	97.4	4.0
Mike Haviland	0	0	711	68.5	4.0
Brent Thompson	0	0	447	42.4	3.9
Ryan Huska	0	0	715	65.4	3.8
Derek Laxdal	0	0	875	78.0	3.7
George Gwozdecky	0	0	956	82.6	3.5
Marty Williamson	0	0	816	66.9	3.4
Marc Habscheid	0	0	842	68.0	3.3
Dale Hunter	60	−7.3	1,028	118.1	3.2
Benoit Groulx	0	0	1,137	87.9	3.2
Darryl Sutter	1,285	49.6	0	0.0	3.2

Minimum 400 games coached.[106]

Are these results meaningful? After I wrote this chapter, the coaches ranked first and sixth on this list were hired by the Philadelphia Flyers and Los Angeles Kings, respectively. Since nobody had seen the results, these two organizations arrived at the same subjective conclusions as this objective analysis. I'd say that passes a preliminary sanity test.

So who's left to hire? At the time of writing, Dan Bylsma tops the list among established coaches, followed by Michel Therrien and former Buffalo assistant coach Terry Murray. It's also fair to say that former NHL coaches like Constantine, Berube, and Johnston deserve another opportunity as well. Among the younger names on this list, we should keep our eyes on Keefe and Bernard now that Knoblauch has been hired. Other off-the-board ideas include Martinson, Malcolm Cameron, and Rand Pecknold.

---

106. I arbitrarily decided that NHL games coached are worth double to include Craig Berube and loosened the constraint on Sheldon Keefe and Martin Bernard, who otherwise just missed the cut-off.

Some of today's most successful teams have noteworthy assistants and/or accomplished former NHL head coaches supporting their head coach. Special teams and/or goaltending coaches have also been known to be integral elements of winning seasons, by helping to improve the team's goal differential that little bit more.

Having devised a reasonable way of measuring the experience and prior success of an individual head coach, the next logical step is to incorporate the contributions of his assistants. The easiest way to do that is to add the head coach's results to those of his assistant coaches. Given that the head coach has more influence over the team's success than his assistants, the latter group should be added in at a lesser weight. For this first pass, I have arbitrarily decided to assign half-weight to a team's assistant coaches. There should also be credit for each season spent as an assistant coach, arbitrarily weighing one full season as being equivalent to 25 games as a head coach.

The results are on the following chart, with the most experienced coaching staffs on the right side and those with the greatest history of prior success at the top. Ideally, a team's coaching staff would be at the top right of the chart.

2017–18 NHL COACHING STAFFS

According to these results, it's reasonable to conclude that the Minnesota Wild and the Dallas Stars have the best coaching staff, based on their prior success. Those who place more value on experience might rank the New York Rangers as the best. Beyond that, teams like Chicago, Ottawa, Pittsburgh, Toronto, and Washington sit atop a pretty tight pack of around a dozen teams.

While this is definitely very interesting, how much can we trust these results? I built charts like these for the 2015–16 and 2016–17 seasons, and they didn't appear to be very predictive, statistically. There was only a slight positive correlation between either experience or prior success and a particular season's results, and the correlation was only a little stronger when we used both together.

Consider the following chart, which places all 60 teams (30 teams over two seasons) on the same chart but replaces the points with sized and shaded circles. The circles are shaded if the team exceeded expectations and white if they didn't, and they're sized based on the extent.

If this evaluation technique were perfect, then all the white circles would be on the left and bottom of the chart and get larger as they approached that corner. Similarly, all the shaded circles would be on the top right of the chart and get larger as they approached that corner. That is roughly what happened, but not to a clear extent.

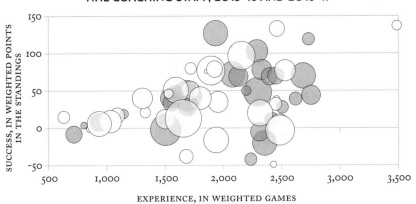

NHL COACHING STAFF, 2015–16 AND 2016–17

Starting at the bottom left end of the chart, many of the teams correctly lack the shaded circles that indicate success. The exceptions are mostly surprise Eastern Conference playoff teams from the 2015–16 season, all of whom missed out in 2016–17. Just like prospects who exceed expectations in the first season tend to struggle in the next, maybe these assessments weren't wrong and it just took another season for the inexperience and/or lack of success to materialize. If so, keep an eye on Jack Adams–winner John Tortorella and the Columbus Blue Jackets, who are the particularly huge success story at 1,500 weighted games. They started the 2017–18 season with the same coaching staff, so by the time this book is published, you'll know if they slid back into the pack or if they're the real deal.

Moving to the top right side of the chart, all but three of the circles are shaded, as expected. The exceptions are the 2016–17 Blues, Kings, and Sabres, each of whom fired their coach either during the season or after it was complete. Unfortunately, that means that we won't get the chance to see if they would have recovered with the same staff.

The particularly large white circle at around 100 weighted points of prior success is the 2015–16 Montreal Canadiens, who lost Carey Price to injury and then spiralled down the standings. The next season, they replaced coach Therrien with Julien and finished first in the Atlantic Division.

So even if Minnesota, Dallas, and the other teams with a strong coaching staff don't exceed expectations this season, it's likely that they will bounce back the following season, although possibly with a different coaching staff.

Closing Thoughts

Coaching is important, but we have a long way to go before we can identify the good coaches from the bad, whether using numbers or the eye test.

Using the difference between a team's results and their expectations is an improvement over judging a coach based solely on his win-loss record, but only a slight one. Plus, it doesn't appear to be very predictive year over year.

Adding in a coach's data from other leagues, like the AHL, CHL,

NCAA, and even ECHL should help, but the weighting assigned to such data was chosen arbitrarily. Likewise, adding in coaching data for the entire staff should be helpful, but that was also weighted arbitrarily. In time, we should establish the most effective weightings.

Even with this admittedly primitive approach, we can produce a rough indication about which teams are likely to have playoff success and which coaches might soon be looking for work. There are even indications about which staff's success may prove to be fleeting.

From this perspective, it seems pretty clear that Dallas and Minnesota have the most accomplished coaching staffs, while the New York Rangers have the most experienced. In Bruce Boudreau, Minnesota may have the world's most accomplished coach, with a track record of great success in the NHL, AHL, and even the ECHL. Even a more sophisticated coaching metric would likely produce similar results.

By combing other leagues, we also identified a handful of coaches who may have promising futures as NHL coaches, including new Philadelphia assistant coach Kris Knoblauch.

These are admittedly modest results, but, in time, more people will work in this area, and we'll devise better metrics that can more reliably measure a coaching staff's impact not just on the standings but on scoring, special teams, and shot-based metrics. I have already had discussions with a few front offices, and though they may only play a very limited role right now, hockey analytics are slowly starting to get involved in coaching selection. Quite frankly, there's too much value at stake for them to be ignored.

I look forward to the day when we have proven evaluation metrics not just for coaches, but for other aspects of an organization, whether that's the scouting department, the medical staff, the front offices, and even the hockey analytics department itself. Until then, this analysis serves as a bold stride in the right direction.

ARE THERE CAREERS IN HOCKEY ANALYTICS?

In 1972, Fred Shero of the Philadelphia Flyers hired the NHL's first full-time assistant coach, Mike Nykoluk.[107] Prior to Nykoluk's hiring, every NHL team had just a single coach standing behind the bench wearing a suit, a fedora, and a grim expression.

The league changed slowly at first, but within a generation or two, every team had an entire coaching staff of four to eight coaches, including an assistant coach to help run the bench, a special teams coach to design and run the power play and the penalty kill, an eye-in-the-sky to watch the game from the press box, a goaltending specialist, a conditioning coach, a development coach, a video coach, a special coaching advisor, and more. Plus, nobody wears fedoras anymore.

It's easy to see where I'm going with this. Most NHL teams today have a massive staff of coaches, doctors, trainers, scouts, equipment managers, and so on, but only a single statistician. Often this single statistician will also have other duties and may not have any particular education or experience in the field. Many of them are paid very modestly, play only a minor role in team decisions, and aren't even acknowledged publicly.

Within a generation, expect that to change. Just like with its

---

107. Eric Duhatschek, "Hockey's Biggest Shift: Fifty Years of Evolution in NHL Coaching," *Globe and Mail*, March 10, 2017, https://beta.theglobeandmail.com/sports/hockey/from-xs-and-os-to-ipads-the-evolution-of-coaching-in-hockey/article34268774/.

coaching staff, each team will have an entire analytics department composed of a variety of specialists. As is the case in a few other sports right now, tomorrow's top hockey teams will scout academic institutions that have the most prestigious mathematics and statistics departments, and they'll be armed with big contracts, looking to hire the best and brightest graduates.

The good news is that the Nykoluk moment has already occurred in this field, back in 2014. In what was termed the summer of analytics, about a dozen prominent hobbyists were hired by NHL front offices in the span of a few months.

At the start of the preceding season (2013–14), Eric Tulsky had measured the level of adoption of hockey analytics, and he could find evidence that only 18 of the NHL's 30 teams were using it at all, mostly in very limited terms.[108] For instance, he could only clearly identify four teams that had a single full-time analytics employee, one of whom also had other duties, and three more who used regular analytics consultants.

After the summer of analytics, in which Tulsky himself was hired by the Carolina Hurricanes, Josh Weissbock repeated this exercise.[109] At the start of the 2015–16 season, he could confirm that 26 of the NHL's 30 teams were using hockey analytics and to a significantly greater extent than before. At that point, there were 16 teams with at least one dedicated statistician, most of those teams had several statisticians on staff, and eight of those teams had built an entire statistics department with a director of hockey analytics that reported to the general manager. Weissbock himself was hired as a prospect consulting specialist by the Florida Panthers soon thereafter.[110]

---

108. Eric T, "Catching Up: The Increasing Pace of Team Adoption of Hockey Analytics," *Outnumbered* (blog), October 21, 2013, http://www.sbnation.com/nhl/2013/10/21/4857618/nhl-stats-advanced-analytics-teams.

109. Terrell Williams, "Hurricanes Hire Eric Tulsky as Hockey Analyst," National Hockey League, August 20, 2015, https://www.nhl.com/hurricanes/news/hurricanes-hire-eric-tulsky-as-hockey-analyst/c-777355; Josh W, "NHL Analytic Teams' State of the Union," *Hockey Graphs* (blog), September 9, 2015, https://hockey-graphs.com/2015/09/09/nhl-analytic-teams-state-of-the-union/.

110. "Florida Panthers Name Richard Pollock Pro Scout and Josh Weissbock Prospect Consulting Specialist," National Hockey League, June 18, 2016, https://www.nhl.com/panthers/news/florida-panthers-name-richard-pollock-pro-scout-and-josh-weissbock-prospect-consulting-specialist/c-886387.

As I write this in the summer of 2017, the adoption of hockey analytics has continued to grow. The NHL launched a new in-house "enhanced stats" platform, all 31 teams employ at least one statistician, leading hockey analytics consulting firm SPORTLOGiQ has contracts with 21 teams, and the Arizona Coyotes even chose statistician John Chayka as their GM.[III]

The growth of hockey analytics is well underway, and many more opportunities will open up in the future. So there *are* careers in hockey analytics, and there will be more. However, who gets these jobs, what do they do, and how do they do it?

Being a long-time pioneer and something of an ambassador in this field, I have been in the unique position to gain a little bit of insight into what's going on behind the scenes. Whether you're interested in a career in hockey analytics yourself or just curious about how these roles are playing out in NHL front offices, I'll do my best to describe what I've seen behind the curtain.

Despite the growing number of opportunities within it, the first step is to think outside the NHL. Everybody's dream is to work for an NHL team, but there are only 31 teams, and many only hire one statistician. Compare that to the average of 500 to 1,000 applications that NHL teams can receive for publicly posted job openings.

Most statisticians get their start outside the NHL. Consider the number of opportunities with the 60 teams in each of the Canadian major juniors and NCAA Division I, plus 132 more teams in the Canadian minor juniors, 36 more in U Sports, and the other 57 North American semi-pro teams between the AHL and the ECHL. Plus, there are also opportunities in women's hockey, the European leagues, and elsewhere.

Furthermore, most NHL teams outsource a lot of their analytics work to third-party agencies, of which there are dozens. In fact, Chayka got his opportunity with the Coyotes by founding his

III. Evan Sporer, "Primer: Get to Know the New Enhanced Stats on NHL.com," National Hockey League, February 20, 2015, https://www.nhl.com/news/primer-get-to-know-new-enhanced-stats-on-nhl-com/c-754260; John Matisz, "What We Learned at the Ottawa Hockey Analytics Conference," *Toronto Sun*, May 8, 2017, http://www.torontosun.com//2017/05/08/what-we-learned-at-the-ottawa-hockey-analytics-conference; Jerry Brown, "Coyotes Name Chayka, 26, General Manager," National Hockey League, May 5, 2016, https://www.nhl.com/news/coyotes-name-chayka-general-manager/c-280574260.

own hockey analytics company in 2010, Stathletes, which now has 50 employees. For many statisticians, careers with one of these companies may prove to be more rewarding than working directly for an NHL organization.

And don't forget that the leagues themselves require statisticians, as do player agencies, and mainstream broadcasters that cover the various leagues and teams. In short, there are a lot of organizations looking for expertise in hockey analytics.

Whether working for an NHL team, in another league, or with a third-party organization, the job opportunities usually boil down to one of three roles: statistical analyst, programmer, or manual tracker. Most opportunities are a combination of the three.

Let's take a closer look at each role, what the job requirements are, what tools and technologies the statisticians use, and how they can help an organization.

Manual Trackers

The first thing any organization needs to do is collect the data, which is why some of the most popular positions right now involve manual tracking. It involves acquiring and classifying NHL data over and above what the official scorekeepers record and gathering even simple data from developmental leagues, which record very little data at all.

These are highly flexible positions that can be either local or remote and either part-time or full-time. They can be with team-level organizations in the NHL or other leagues or with a growing number of third-party companies to whom this work is commonly outsourced. Either way, here's how they work.

1. THE TRACKER WILL BE SENT VIDEO OF THE GAMES

Unless the information is being compiled by official NHL scorekeepers or is being used to report directly to the coaching staff at intermission or immediately after the game, the trackers are not required to attend the games and do the work live. Tracking live is exceptionally difficult to do quickly and accurately, unless the events being tracked are few and simple, the tracker is using special technology, or the tracker is particularly experienced.

More commonly, video for multiple games is sent to the tracker, who will complete the assignment on evenings and weekends, usually taking a few hours per game.

▍▍ 2. THE ORGANIZATION WILL SEND THE
▍▍ TRACKER A LIST OF EVENTS TO TRACK

The most common requirement is to track shot attempts, followed by puck touches, passes, or zone entries and exits. However, I have heard of at least 50 different event types over the years, including screens, stretch passes, plays up the end boards, hits that resulted in a turnover, and pretty much anything else you can think of.

Tracking the most common events is usually pretty straightforward and requires very little experience to complete accurately. However, some events require considerable hockey knowledge to properly interpret and record, and they are therefore usually assigned to former players, coaches, and scouts, along with careful instructions about how these events are defined and how to record them.

▍▍ 3. THE ORGANIZATION WILL SEND THE ANALYST SOME
▍▍ SOFTWARE USED FOR VIDEO TAGGING OR PERHAPS A LAPTOP
▍▍ WITH THE REQUIRED SOFTWARE ALREADY INSTALLED

In some cases, the organization wants more than just a count of the events and who was on the ice. They want an actual video file tagged with the time of each event. That allows the coaches to quickly study specific plays and matchups.

Video tagging requires a little bit of training and specialized technology. Leading examples of such software include Steva Sports Software (stevasports.com), XOS Digital (xosdigital.com), Gamebreaker (Sportscode) by Digital Tec Solutions (dtsvideo.com /gamebreaker-plus), PUCKS by Sydex Sports Software (sydexsports .com/html/hockey.html), and dozens of others.[112] Licences are quite

---

112. At the time of writing, TVTI, which was the video provider for PUCKS, had been bought by Stats LLC, which is making their own version of PUCKS called STATS. PUCKS now has a new video provider.

expensive and possibly unavailable to those outside NHL organizations, so the software must be provided by the team. To my knowledge, there is no industry standard for video tagging, and some organizations use their own proprietary software.

Depending on the exact technology, this software allows the trackers to watch a video of the game and to quickly tap the screen to record the type of event, along with additional information, such as the time it occurred, where it occurred on the ice, and the jersey numbers of the players involved. It's quite remarkable to watch how quickly and accurately a seasoned expert can tag a game using sophisticated software like this.

Unless video tagging is required, most manual trackers will do the work by hand either with pen and paper, directly into a spreadsheet, or by using custom software developed for just this purpose.

▉▉▎ 4. THE TRACKED DATA IS RETURNED TO THE ORGANIZATION

The end result is usually a formatted list of times, events, locations, and jersey numbers that is produced manually or is exported from the software in a variety of formats. If video tagging was required, then the end results will include a tagged version of the video itself.

If this was done for an NHL game, then the results can be integrated into the league's official shift charts and play-by-play files.

And that's about it.

Since the work is flexible and requires no experience in statistics or programming, it is a great first step into the world of analytics. In time, the manual tracker can be involved in the subsequent analysis of this data and/or the programming tasks required to store it, manipulate it, and create reports.

To get started in a career in manual tracking, the first step is to find (or create) and get comfortable with a convenient process for tracking game data quickly and effectively. This is often built into a spreadsheet, but there are a number of cheap applications out there. Manual tracking can even be done with XOS, and it pays to learn the video tagging technologies anyway.

Next, learn how to track as many different kinds of events as possible. Even the simple events aren't as straightforward to record

-REFS IN THE OFFSEASON

CHECKING   BOARDING   HOOKING   CHARGING   ICING

---

as most people would expect, since hockey has a way of creating so many wild situations.

Try to get some experience by contacting local junior, university or college, and minor-league teams in your area, even offering to work for a low wage or on a volunteer basis. Another way to get your feet wet is to consider joining an online community that does manual tracking for independent research purposes, like Ryan Stimson's Passing Project.[113]

Even if it doesn't net you the big bucks, this experience will help prepare you for a meeting or an interview with solid examples of your work. I also recommend measuring the speed and accuracy with which you record various types of events and how fast you can tag them using the different types of software.

When looking for job openings, don't just look at the team or league level. In many cases, the work will be outsourced to third-party organizations. One of the more famous cases is Stathletes (stathletes.com), based out of St. Catharine's, Ontario. They were co-founded by John Chayka, who went on to become the youngest NHL GM in history, with the Arizona Coyotes.

Hockeytech (hockeytech.com) is another big player. They were founded by former Panthers CEO Stu Siegel in 2013, based on corporate entities that go back to 1998. They have a variety of services

---

113. You can support Stimson's Passing Project at his Go Fund Me page, www.gofundme.com/between-the-lines.

and clients in virtually every North American league, including every NHL team.

And then there's PowerScout Hockey, founded in Ottawa by Marc Appleby. They were one of the earlier manual tracking companies known to work in the NHL and were a pretty big player for years. However, I haven't heard much about them since 2014, and their website is long gone.

There are also a number of small manual tracking companies launched by former NHL players. Two good examples include Brad Werenka's Calgary-based Truperformance (truperformancehockey. com) and Drake Berehowsky's StatsTrack (statstrack.ca) in Toronto. If you live in or near an NHL city, check around to see if there are opportunities to learn and contribute.

If not, many prominent hobbyists have chosen to co-found their own manual tracking company, including Jennifer Lute Costella and LCG Analytics (lcganalytics.com) and Garret Hohl, whose Vancouver-based company HockeyData (hockeydata.com), which focuses on the AHL, landed a contract with the Washington Capitals in May, 2017.[114]

And, the list goes on. There's also Bench Metrics (benchmetricshockey.com), Dark Horse Analytics (darkhorseanalytics.com), and InStat (instatsport.com/en). There could be as many as 100 of these companies. Almost without exception, they also offer analysis and reporting software and services, so they should therefore be of interest to statisticians and programmers, too.

Lastly, manual tracking isn't the only way that teams are getting their data. In the future, video technology will play a larger role. Due to the cost of the special equipment and software, video is far more expensive than manual tracking. However, once it is automated, additional games and countless events will be tracked at virtually no additional cost, and accurate results will be ready within seconds. If enough data is being tracked, or the expense is being shared by enough organizations, then video technology will eventually be as affordable as manual tracking.

---

114. Randy Shore, "Vancouver Entrepreneurs Net Analytics Contract with NHL's Capitals," *Vancouver Sun*, May 10, 2017, http://vancouversun.com/news/local-news/vancouver-entrepreneurs-net-analytics-contract-with-nhls-capitals.

There are a number of companies who have been introducing this technology to hockey, including PowerScout Hockey, SPORTLOGiQ (sportlogiq.com), Iceberg Hockey (http://iceberg .hockey), Keemotion (keemotion.com), Pixellot (pixellot.tv), StriveTV (strivenetwork.tv), and Hawkeye (hawkeyeinnovations. com), which is also used for coaches' challenges in the NHL.

As the technology evolves, video tracking can also improve the accuracy of the data. That trend has already been observed in other sports, like with PITCHf/x and FIELDf/x in baseball and SportVU in the NBA. Even when multiple special cameras are installed, hockey presents a few interesting challenges for video recognition, like when players get tangled up in a puck battle or when the puck gets obscured by the boards, a player, an official, or the net. That's where augmenting video camera technology with chips in the pucks and/or equipment can come in handy, albeit at even greater expense.

As observed at the 2015 NHL All-Star Game and the 2016 World Cup, using Sportvision technology in the pucks and player tags can provide precise location information several times per second and can even track the angle and speed of the players and puck. Interestingly, hockey was the very first sport to use this technology, which was initially known as FoxTrax (with the infamous glowing puck), all the way back in 1996. Acquired by SMT in 2016, hockey is now listed under the "other sports" category on the company website (sportvision.com).

Despite the growing success of these and other technologies, they are unlikely to ever drive manual trackers out of business. There will always be many leagues for which it just isn't cost-effective to pursue those options, and there will always be specialized events that temporarily fall outside the bounds of what a camera or computer can offer.

Data Scientists

The role of the data scientist is the easiest to describe, because it's very close to what we do in books like these. We start with a question, form a reasonable hypothesis, carefully gather the data, craft and/or identify appropriate measurements, devise tests for our hypothesis, calculate the results, interpret them, and then express the results in a way that makes sense to everybody. In a nutshell, that's exactly what a data scientist does for a hockey team.

Surprisingly, only a few teams invest in data scientists. The others are content to outsource this work or to hire an entry-level analyst who creates reports that merely apply the analysis done in the public sphere.

In time, more organizations who are looking for an edge will hire and retain statisticians and other innovative thinkers of their own. Over the years, we have actually been sneaking some of the more sophisticated terms and techniques used by statisticians into these books, especially in the opening chapter of *Stat Shot*. Long-time readers might actually understand this world better than they realize.

For example, let's consider a case where the statistician's input is being requested for a specific draft pick, a trade, or a contractual decision. The statistician might be a full-time employee, part of a third-party organization, or perhaps just a consultant being engaged on this one matter.

Let's imagine that it's the summer of 2010, you're a statistician for the Montreal Canadiens, and the team's new GM Pierre Gauthier asks you if he should keep Carey Price, who had a career save percentage of 0.912 at that point of his career, over Jaroslav Halak, who had 0.919.

A hobbyist would buy the latest version of *Hockey Abstract*, download the latest data from a few websites, apply what they have learned, and create a good report for Gauthier. There's nothing wrong with that (especially the part about buying the latest version of *Hockey Abstract*), since that contribution would certainly be of good value to the organization. However, statisticians can take it a step further by innovating custom metrics and producing results of greater statistical certainty.

It all starts with tests. Statisticians run a lot of tests, which is exactly how this hypothetical situation would first be approached. Here, the statistician could test the hypothesis that Halak's superior skill resulted in the higher save percentage versus the so-called null hypothesis that it did not.

Of course, statisticians never deal in absolutes. Statistical analysis is all about probabilities, not certainties. The statistician would have to select a level of uncertainty that is acceptable, which is called the alpha level. Typically, this is 0.05 or 5%.

Establishing that level of uncertainty means calculating the p-value, which is the probability that the null hypothesis is true. In other words, it is the probability that the observed difference in save percentage could have occurred without Halak possessing greater skill. Generating that p-value usually means conducting a sample test of the data, determining the degrees of freedom, calculating the standard error rate, and building a distribution curve. Explaining these terms is outside the scope of this book but don't worry, there won't be a test on any of this.

If that p-value is less than the chosen alpha, then the statistician can reject the null hypothesis and report to Gauthier with 95% confidence that Halak is better than Price. If this later turns out to be one of those 1-in-20 cases where the numbers were wrong, then that is termed a Type I error. When getting into an argument with a statistician on Twitter, I highly recommend you throw that term at them, since it really gets their goat.

In hockey analytics, Type II errors are far more common. Those occur when the p-value exceeds alpha, which means the null hypothesis can't be rejected, even when it is false. In our example, it means that the data didn't definitely indicate that Halak was better when he actually was. While that obviously didn't occur in this particular case, it often happens when there isn't enough data or the data just isn't good enough.

For example, in *Stat Shot* we found that "Braden Holtby, in second place, has a lower bound that exceeds the upper bounds of only two goalies, journeyman backup Curtis McElhinney and Jonas Gustavsson."[115]

That means that we couldn't reject the null hypothesis when comparing Holtby's career even-strength save percentage to almost

---

115. Rob Vollman, "Who Is the Best Puck Stopper?", *Stat Shot*, 2016, 178–179.

any active NHL goalie. Arguing that factors other than skill can produce the difference between Holtby's 0.929 save percentage and Ondrej Pavelec's 0.917 is a great example of a Type II error.

If it sounds complicated, well, it is. Plus, statisticians wouldn't even be using something as simplistic as save percentage in a situation like this. Instead, they would be using their own data sets, and multivariate metrics of their own design to measure the meaningful and persistent qualities of a goalie's skill, and then they would use them to compare Price and Halak not just against each other, but against all goalies. In fact, that's exactly where we landed at the end of that goaltending chapter in *Stat Shot*, with a rather comprehensive and sophisticated model of all goalies.

Furthermore, that statistician certainly wouldn't go to Gauthier with talk of p-values and Type II errors, just as we generally don't include those terms in books like these. These books are mostly written for the enjoyment of mainstream fans and hobbyists, and the professional statisticians among you have simply learned to tolerate my oversimplifications (including those previously mentioned). Reports for management have to be filtered and crafted similarly.

At first, most data scientists do all of their work in Microsoft Excel because it is popular, is easy to use, and requires no experience in either programming or statistics to learn and use, unless it is supplemented by an extension like Visual Basic or SigmaXL. According to a survey conducted by Ryan Stimson, 73% of the field use Excel as their primary tool.[116] Even if most statisticians eventually move on to more sophisticated technologies, Excel will always remain a useful technology with which to share one's findings with the rest of the community and/or the front office.

Second place in Stimson's survey went to R, which is gradually becoming the standard in this field. I highly recommend it for hobbyists who want to take their work to the next level, whether or not they are seeking a career in hockey analytics.

What is R? Besides being a very nifty letter with which to spell first names, R is a programming language. Rather than clicking on

116. Ryan Stimson, "Hockey Analytics Survey Results," *Hockey Graphs* (blog), February 9, 2017, https://hockey-graphs.com/2017/02/09/hockey-analytics-survey-results/.

Excel cells, statisticians do their work by typing commands into the R console and grouping sets of commands into programs. It has a bit of a learning curve for those without any programming expertise, but pretty soon statisticians are able to complete the same tasks they were doing in Excel in a fraction of the time and with minimal effort.

If you're interested in doing a little bit more with statistics, then get started by downloading and installing R, which is free. [117] It comes with the RGui console, but you may want to download and install RStudio, which has a graphical user interface (GUI) that's easier to use and has a few more features.[118]

As for data, R can actually open all of your Excel files as well as any files in the formats used by other statistical tools and technologies. If you're just starting out, then you can grab the data from my website (hockeyabstract.com/testimonials). Load the data using the library command, and then use the read command to open it. Next, start by using the head and summary commands to make sure everything's working and to get your first look at the data set in question.

To start using and analyzing the data, there's no need to be an experienced programmer, since one of the great advantages of R is having access to a lot of great libraries and packages that can manipulate the data, do all the number crunching, and create visualizations for you. For starters, download Michael Lopez's list of the 10 most useful R packages for sports statisticians, to which I would also add NHLScrapR to parse NHL game files.[119] Also, if you're going to be in the public space, then it can be helpful to create a Github account of your own (github.com), in order to organize and share all of their code and solicit assistance and advice from others.

While R is gradually becoming the standard, there are many alternatives. For instance, APL (or J) and MATLAB are very similar statistical programming languages. However, the former is quite

---

117. The R programming language can be downloaded at www.r-project.org/.

118. RStudio can be downloaded at www.rstudio.com/products/rstudio/download/.

119. Michael Lopez (@StatsbyLopez), "Here's a list of #rstats packages, catered to those doing statistics in sports," Twitter, November 8, 2016, 7:38 a.m., https://twitter.com/StatsbyLopez/status/796014026094481408; the NHLScrapR package can be downloaded from the War on Ice Github page, https://github.com/war-on-ice/nhlscrapr.

antiquated, and the latter can be expensive and is more common in academia or other fields. They will absolutely do the job and are a huge step up from Excel, but using these tools makes it a bit more difficult to collaborate with others.

Since it's impossible to know which tool or technology a team is using—or their third-party partners or various independent consultants—collaborating with others might mean having all the options installed on your laptop and acquiring a basic working knowledge of each one. In addition to R, APL, and MATLAB, that might include:

- Minitab, which is extremely simple and therefore popular in academia for instructional purposes. Nobody really uses it for their day-to-day work, but everybody knows it.
- SPSS, which is IBM's highly sophisticated analytics software whose popularity is stunted only by its expense. For those who can afford it, it's the most popular alternative to R, in my experience.
- SAS, or Statistical Analysis System, which is a collection of software tools used for a wide variety of applications in the field of math, data, and statistics.
- Stata, which is very established statistical software that is far more popular in other fields than in sports, much like MATLAB.
- Gretl is a free open-source statistical package, with a GUI that can actually be used with other languages, like R and Stata.
- SAP technology, which is an enterprise resource planning software with an expensive and complex analytics solution. It is used by the NHL and virtually nobody else in this field.

Data scientists may also have a variety of more general-purpose programming languages on their laptop, like Python and SQL, which we'll discuss in the next section.

Given that analytics is used extensively in other fields, such as sales and marketing or business development, you may already have access to some standard tools and technologies, including the licences to use them. To be clear, most organizations don't have any overlap between the data analysts working in their different departments, but there's no reason that they can't share the technical infrastructure and IT staff.

Regardless of which technology you use, accuracy is essential, especially in data collection. Even if the data is acquired from third parties or manually collected by other people in the organization, the data scientist has the responsibility to organize these massive data sets, to verify their accuracy, to identify and remove any bias, and to become thoroughly familiar with every scrap of information, including their relationship with the scoring and preventing of goals and their year-to-year persistence.

Sometimes data scientists are criticized because, in some people's minds, only wins and losses matter, so we should just toss all the fancy numbers away. However, wins and losses being the only thing that matters is actually the foundational argument of the entire field. In fact, it is the first law of hockey analytics.[120] If a fancy number can't be shown to have a reproducible relationship with wins and losses, then it is tossed aside. That's the entire point of the field, making that criticism one of the field's more bizarre ones.

Misunderstandings like these create the need to communicate and express the results in a meaningful and actionable fashion, whether that's in reports, emails, or presentations. This is especially true in these early days, when there is not likely to be as good a foundational understanding of hockey analytics throughout the organization nor a common language. In many organizations, there's also a degree of skepticism that can go well beyond what is healthy.

As for experience playing hockey, that isn't strictly required because most organizations already have that in spades. However, statistical data can be highly prone to misinterpretation, and it pays to have a solid grounding in how the sport is played.

As for pay, statisticians average about $85,000 per year in the United States and about 50% more in specific sectors or particular geographic locations. In hockey, they may make only half as much. Nobody goes into this field for money or fame; there isn't very much of either one.

---

120. Alan Ryder, "The Ten Laws of Hockey Analytics," Hockey Analytics, January 2008, http://hockeyanalytics.com/2008/01/the-ten-laws-of-hockey-analytics/.

Surprisingly, those who have been most successful at finding work in hockey analytics right now are those with programming skills.

Dating back to Darryl Metcalf, who built the Extra Skater website before being hired by the Toronto Maple Leafs on August 19, 2014, those who can build databases of the field's leading research are perceived to have equal or greater value than those who created such stats in the first place.[121] Even a long-time brilliant hockey statistician like Andrew Thomas wasn't hired until he built the War on Ice website to share his and the rest of the field's latest developments with the general public.

Surprisingly, the programming positions are the ones that I know best, since that's what I used to do for a living. Prior to doing hockey analytics full-time, I spent 15 years in a variety of fields doing programming, technical support, and project management for large data-driven applications. This background has also given me particularly keen insights into what these programming positions entail and what teams are looking for.

The primary responsibilities for a programmer in a hockey analytics department boil down to the acquisition, storage, manipulation, and retrieval of data. In some cases, these are full-time, on-site positions, and in other cases they are done part-time and remotely, or they are outsourced to another company entirely.

Every hockey organization will eventually build a database of hockey stats. Virtually all organizations will already have databases for their other business departments, like human resources, accounting, marketing, and so on. Often, the hockey analytics department will share the same infrastructure, underlying IT staff, and software licences as the rest of the organization.

Building that database and populating it with information will be one of the programmer's key responsibilities. The biggest challenge is that the data will come in a variety of different formats,

121. James Mirtle, "Leafs Go Big on Analytics with Creation of Stats Department," *Globe and Mail*, August 19, 2014, https://beta.theglobeandmail.com/sports/hockey/leafs-go-big-on-analytics-with-creation-of-stats-department/article20129261.

depending on whether it comes from the league, is scraped from the websites mentioned throughout this book, is acquired from third-party data providers, or is manually gathered by the team itself. All of this information must be acquired in an automated fashion and stored in the database in a common format.

Even once the data is converted to a common format, there's often a lot of work that needs to be done to clean it up. That means looking for missing or incorrect data or pieces of data that conflict with others. Just to take one simple example, there might be one data source that lists a player as Matt Benning while another lists him as Matthew Benning. Sometimes that change will take place in the middle of the season, and there will be two data sets for a single player that need to be merged.

Programmers also want to avoid having redundant information in the database. What if they don't add up? For example, goals and shots will be stored in the database, but shooting percentage will not because it is simply a different representation of the other two statistics. Having all three pieces of information in the database could lead to some sticky situations, trying to figure which data is accurate whenever shooting percentage doesn't exactly match goals divided by shots.

That's one situation where having at least a basic understanding of the field is handy. Shooting percentage may be a simple enough statistic for everybody to wrap their walnuts around, but what about GVT or expected goals? Even if the team has a few data scientists to help out, the programmer will still need to have a working knowledge of stats and hockey in order to program the various calculations properly.

Finally, there's the task of making all of this wonderful information available to the rest of the organization. Programmers can build websites like Extra Skater and War on Ice, and they can also produce automated reports that are sent directly to the laptops, browsers, phones, or tablets of coaches, scouts, and GMs. New features are requested all the time, but the programmer is often given a lot of freedom as to how it's done.

To complete these tasks, programmers use a number of tools. To highlight just the four most common ones:

- The data is normally acquired using technology like Python, which is a scripting language ideal for scraping data off the internet and manipulating it for further analysis.
- There are alternatives, like C#, which is the spiritual successor of the technologies I used to use. It is a generic but very popular and powerful programming language that can build almost anything from scratch.
- To store the data into the database, programmers use SQL, which is the Structured Query Language used in databases like Microsoft SQL Server to store, manipulate, and retrieve data of any kind. I used this technology quite extensively, but for a database called Oracle. I even had a very popular website in my previous career (thinkoracle.blogspot.com, which is still online) and was the technical editor of one of the best-selling Oracle SQL textbooks of the time (*Oracle SQL by Example* by Alice Rischert).
- To present the data, there's Tableau, which is arguably the best data visualization software out there for making charts and graphs.

As for compensation, if you thought the pay cut that data scientists have to take in order to work in hockey is extreme, then you really don't want to know about the programmers. The good news is that the flexibility inherent in these positions often means that programmers can work from home and on their own schedule.

Career Advice

Regardless of which position you may want to pursue, my best advice is to learn the fundamentals, don't be afraid, find and trust your voice, and stay focused on your passions.

In the past, my own sense of modesty caused me to resist requests for advice—who the heck am I? But as the field grew and everybody's successes began to accumulate, I started putting a little bit more thought into which of our experiences might be of value to those who are just getting into hockey analytics as a possible career.

At first, my approach was to identify and repeat the best advice that I had received, rather than come up with anything of my own. As I wrote in *Hockey Prospectus* just prior to the 2014–15 season, the best advice I ever received came from a rather unexpected source: a stand-up comedian named Darryl Lenox. His amateur comedy workshops taught local comics how to find and trust their voice, how to keep coming up with good ideas, and how to build their audience.[122]

Even though Lenox's advice was geared toward the stage, it translated well to my glorified hockey stats hobby, too. Rather than try to emulate other stand-up comedians that we admired, his most important piece of advice was to find our own voice and stick to it. The sooner you do that, he argued, the less time you waste developing the wrong talents for someone else's audience.

In our world, that means that we can be inspired by hockey statisticians whose work we admire, but we need to find our own way of approaching this game. We need to find the specific areas that interest us and what approach we'd like to take. Here are 12 questions that can help you figure that out:

- Is this meant to be a casual hobby, or do you want to work in a front office or in mainstream media or elsewhere?
- Do you want to stick to writing, or do you want to make appearances on radio, TV shows, and podcasts?
- If you're a writer, are you interested in blogs, formal online articles, magazines, or books?
- Is your intended audience coaches and front offices, the mainstream fan, the fantasy pool enthusiast, the hard-core statistician, or someone else?

---

122. Rob Vollman, "So You Want to Be a Hockey Analytics Writer?," *Hockey Prospectus* (blog), September 5, 2014, http://www.hockeyprospectus.com/so-you-want-to-be-a-hockey-analytics-writer/.

- Is your natural style to muck it up and argue against bad ideas, or to shine a light on the great ones?
- To what extent do you prefer tables of figures, visualizations, or the written word?
- Do you see yourself developing new statistics and innovations, or finding applications and/or new presentations for what's already out there?
- Can you write code to parse data from websites and game files and create automated reports?
- Do you want to create and host your own statistical database website?
- Do you have any interest in recording your own statistics, like zone entries or passing data?
- Is there a specific team you want to cover, or do you want to cover the NHL as a whole or perhaps specialize in another league altogether?
- Do you want to cover the entire game, or do you want to focus on an individual area, such as goaltending, faceoffs, injuries, or special teams?

Even once these questions are answered, it can still take time to find and refine that voice. In my case, I spent an invaluable 2013–14 season writing for Bleacher Report and investing time reading all of the comments to figure out how best to phrase my ideas for the average fan, including those who were new to and skeptical of statistical hockey analysis.

All that interaction with my readers also helped build my readership. Lenox described comedy as a competitive and individual business, not a collaborative one. Sure there are lots of people who will support you and help refine and promote your work, but ultimately a comedian will always find himself alone on a stage with nothing but a microphone, trying to be funnier than the next guy in order to land his next gig. Likewise, in hockey analytics, it will be your name at the top of an article, with only the quality of your own analysis helping you score the next opportunity.

To that end, Lenox's advice was to create your own market.

When he comes to town, the comedy club already has an audience and books him only because of the additional people he will bring in. If he doesn't bring his own audience, then they will book someone else who can.

Don't rely on a website, radio station, front office, or publication to promote your work or to provide the audience for you. Cultivate it yourself. Reply to emails, respond to tweets, speak at conferences, go to games, and get engaged in the comments sections and newsboards. This is invaluable information that will help you discover who your audience is and how to hit their mark.

There will be a lot of criticism—from hockey traditionalists, from fans, from the media, and most especially from your colleagues within the hockey analytics community itself. No matter what you do, there will always be about five to ten critics for every supporter. Always remember that the only way negative people can be right about you is if you quit.

Finally, Lenox's closing advice was to write every day and the ideas will surface. There is absolutely no substitute for hard work, and you'll need a genuine passion for this field in order to stick with it.

Although it was meant for a completely unrelated field, Lenox's advice got me a long way in hockey analytics. Since then, I've been involved in a lot of hiring decisions, and I've picked up the following more practical and specific pieces of advice directly from those in front offices:

▰▮ 1. CREATE SOMETHING

Whether you write a paper or an online article, hold a conference, invent a statistic, build a website, found a company, participate in a manual tracking project, or start a podcast, it helps to create a name and an identity for yourself.

▰▮ 2. NETWORK AND BUILD UP YOUR CONTACTS

Everybody can teach you something valuable if you have an open mind and listen. With regard to jobs, most of them are not posted publicly, and those that are generally receive between 500 and 1,000 responses, depending on the market. Either way, those who have contacts are those who hear about these opportunities first and get on the short lists.

■■I 3. GET SOME WORK EXPERIENCE

Contact the junior, college, and minor-league teams in your area and offer your assistance. Think outside the box, and contact player agencies and third-party consulting companies. Even if most of them ignore you, or even if it's part-time or unpaid work, you only need one response to get started.

■■I 4.TAKE YOURSELF TO THE NEXT LEVEL

Whether you're into manual tracking, programming, or statistical analysis, take the time to master your craft. That could mean studying textbooks, using software and other technology, or getting lots of practice in order to complete a wider variety of tasks or to complete them faster and with greater accuracy (just like an NHL player would).

■■I 5. BUILD A PORTFOLIO

When you get that meeting and/or interview, it helps to be prepared with examples of your work, even if it's just in one small area of the game. Make it something practical and memorable, and tailor it to the organization in question, if possible.

■■I 6. GET YOUR FINANCIAL HOUSE IN ORDER

Quite frankly, it's shocking how little most of these jobs pay compared to similar work in other fields. This is a by-product of how badly so many people want to work in hockey and how many organizations aren't yet properly budgeted for analytics. It would be highly unfortunate to pass on a great opportunity because of debts, a high cost of living, or the inability to relocate.

■■I 7. DON'T BE AN ASS ON SOCIAL MEDIA

I have personally seen some golden opportunities get flushed down the toilet because someone was acting highly unprofessionally and disrespectfully on social media, usually by trashing players, front offices, journalists, or fellow statisticians. Most of the time they didn't even realize that they had blown a great opportunity. And for goodness sake, do not write about sex, religion, money, or politics, the last of which appears to be the hardest to resist!

▥ 8. CONFIDENTIALITY IS KEY

The NHL is understandably a very competitive and secretive league, so discretion is imperative. It can be hard to resist shouting it from the rooftops when you've done something amazing, but most teams simply can't risk working with anyone who might leak information. It's important to have (and keep) a good reputation in a relatively small field where most people keep in semi-regular contact.

▥ 9. SHOW, DON'T TELL

Whenever you see a statistician fail to make a point about Corsi or PDO, it's probably because they're telling others how it works rather than showing them. Part of the appeal of player usage charts is that we're not telling the reader about zone start percentages, of quality of competition, or usage, or Corsi—we're showing them. Likewise, don't go to a meeting or interview to tell someone what you can do for their organization; go prepared to show them. That might mean video or a demonstration or a chart or some tables or specific recommendations—who knows? But find specific ways to show them how you can help.

▥ 10. BE PREPARED FOR SACRIFICE

Remember: the passions of even the fortunate few who were offered front-office positions went unrewarded for the longest time. And even when they landed their first great opportunity, they were asked to shut down their life's work, quit their (generally) lucrative day jobs, and toil for an average of 66 hours per week in potentially a very limited corner of the organization.[123] That's a sacrifice that only the truly devoted can make.

---

123. Michael Lopez, "What's It Like to Work in Sports Analytics?," *Stats by Lopez* (blog), January 15, 2017, https://statsbylopez.com/2017/01/15/whats-it-like-to-work-in-sports-analytics/.

Despite the premium placed on discretion, my involvement in more than a few of the recent hires has been a rather poorly kept secret. Without confirming any of the specifics, I can explain how this came to be.

It started about four years ago. It might surprise you that NHL teams had difficulty finding and identifying good analysts, but that was the case at the time, and to an extent remains so today. It's not like there's a minor-league system for statisticians or a lot of schools or recruiting firms that specialize in hockey analytics.

As one of the more visible people in this field, I started getting a lot of phone calls from various front offices to get opinions on certain individuals or to see if I knew anybody who could help them. At first, I put them in touch with people I knew, including those whose work had been featured in the original *Hockey Abstract*, but I ran out of friends pretty quickly. Those who weren't snapped up in the 2014 summer of analytics were certainly gone by the end of 2015. Plus, there were so many talented new people arriving to the scene that I couldn't even be confident that I was actually presenting these organizations all of their best options.

That's why I opened things up in February 2016 and invited everybody who might be looking for a future in hockey analytics to contact me and to let me know what they're looking for.[124] Then, I could pass along the right opportunities to the right people and give front offices even more options from which to choose.

As you would expect, the response was a little overwhelming, and hundreds of people reached out to me. Some people were just looking for a little bit of advice and encouragement, or maybe just a little part-time hobby work on the side, but others were deeply passionate about full-time careers in the sport.

It took me nine months, but after taking up to a dozen calls per week, I managed to follow up with all the initial respondents and helped a lot more people get started. Several dozen got audiences

---

124. Rob Vollman, "Do You Want an NHL Job in Hockey Analytics," Hockey Abstract, February 9, 2016, http://www.hockeyabstract.com/thoughts/dowantannhljobinhockeyanalytics.

with NHL front offices, and several of them got full-time jobs this season. Of course, during the time I wrote this book, my backlog swelled to an even greater size than before. Once more into the breach!

People ask me why I do this, but it was a highly rewarding and educational experience for me. I was energized from meeting so many bright and passionate young people, and I learned about some of the exciting projects they're working on. I also got the opportunity to learn a lot about how teams are building their analytics departments, what kinds of opportunities are available for those in our field, what skills teams are looking for, and what people have done to get ahead, all of which I've shared here.

Ultimately, I think our careers in hockey analytics are whatever we create for ourselves. When I started, there were no hockey analytics books, no hockey stat websites, and no statisticians discussing our sport in newspapers, online, on TV, or on radio. There were virtually no statisticians working in front offices, no hockey analytics conferences, no visualizations, and very few stats beyond goals, assists, and plus/minus. We had to create all of these opportunities from scratch.

If anything, front offices and other hockey organizations have their ears and eyes open for us to tell them what opportunities should exist and how to create them. If you wait until an opportunity exists before you go for it, then you might just miss it.

QUESTIONS AND ANSWERS

One of my favourite applications of hockey analytics is to answer questions. These are often topical questions, like who won the Shea Weber for P.K. Subban trade, but others are timeless enough to include here.

Usually, the demonstration of how stats can be used is more interesting than the answers themselves. However, that all depends on the question. In this book, we'll look at the value of a penalty shot versus a power play, when teams should pull the goalie, if Ovechkin will ever catch Gretzky, how the NHL can boost scoring, and what might be the key individual player statistic. Let's begin.

Which Is Better, a Penalty Shot or a Power Play?

This is more of a theoretical question, since neither team has any control over whether an infraction will be a penalty shot or a penalty, but it is an interesting question nonetheless.

A penalty shot is obviously better than a 5-on-4 power play. Going all the way back to the 1934–35 season, NHL players have scored on 565 of 1,561 penalty shots, which is 36.2%.[125] Going back to just the 2005–06 season, it's 246 goals in 745 shots, which is 33.0%. That's far better than a 5-on-4 power play, which has ranged between 17.3% and 19.1% over this time span and carries the risk of a short-handed

---

125. Shootout data from the National Hockey League, accessed August 1, 2017, http://www.nhl.com.

goal. So if the play is currently at 5-on-5 or is about to be, then the penalty shot is a better result.

What if the team is already on a 5-on-4 power play? Since 2005–06, there have been 1,731 goals at 5-on-3 in 3,941 advantages, which is a 43.9% success rate. Or to put it another way, it's 25.7% higher than a 5-on-4. So even if a team already has a 5-on-4 power play, it's slightly better off with a penalty shot than a 5-on-3 power play.

The advantage could change, depending on who is assigned the penalty shot and which goalie he is facing. If someone like Marc-André Fleury of the Vegas Golden Knights is in nets, who has a career 0.740 shootout save percentage in 269 shots faced, then a team might be better off with the 5-on-3 power play. On the other hand, if the shooter is someone like Washington's T.J. Oshie, who has scored on 54.9% of his 71 career shootout attempts, then a penalty shot remains the far better situation.

Time is also a key element. If a team is in the lead and wants to kill the clock, then maybe it's better to have those extra two minutes with the man advantage. Even with a 5-on-3 advantage, scoring a goal will end the first penalty, allowing the team to continue to play 5-on-4 for the duration of the second one. Even once all of the penalties expire, this power play may have triggered a momentum change and/or tired out the top opposing players.

Furthermore, these calculations were all based on 5-on-3 power plays, including those that lasted mere seconds before the first penalty expired and those scheduled for a full two minutes. If the 5-on-3 is for the full two minutes, then it could boost a team's chance of scoring by far more than 25.7% and possibly more than 33.0%.

As Gabriel Desjardins charted, a team's chance of scoring on a 5-on-3 that will last less than a minute is below 30% and doesn't cross 50% until the duration is at least 1:20.[126] With so few 5-on-3 penalties of longer duration, it's not easy to extrapolate what will happen after that, but it does mean that there's probably a crossover point beyond which the 5-on-3 will last long enough that a team is better off having a 5-on-3 power play than a penalty shot.

126. Gabriel Desjardins, "Scoring Probability vs Length of a 2-Man Advantage," *Behind the Net* (blog), December 9, 2008, http://behindthenet.ca/blog/2008/12/scoring-probability-vs-length-of-2-man.html.

As a general rule, teams are better off with a penalty shot. However, given the shooter/goalie combination, the game situation, the respective abilities of the special teams, and the duration of the 5-on-3, there will certainly be exceptions.

When Should Teams Pull the Goalie?

The *New York Times* has a Fourth Down Bot for the NFL (nyt4thdownbot.com), which calculates what the offensive team should do on fourth down: punt, kick a field goal, or go for it. It's a purely statistical calculation to maximize the points scored, based on probabilities. In practice, they have found that NFL coaches are far too conservative.

The same conclusion would be reached if there was an equivalent tool for the NHL called the Pull the Goalie Bot, which would calculate the exact time when the trailing team should pull its goalie.[127] Again, it would be based on a simple calculation meant to maximize the chances of tying the game.

The conventional wisdom is that teams trailing by a goal should pull their goalie with around one minute remaining, but this is way too late, statistically speaking. Thanks to early hockey statistician Donald G. Morrison, we have known since 1976 that the optimal time to pull the goalie is with at least two and a half minutes remaining in the game.[128] This result has been revisited and confirmed in a variety of studies since then, culminating in the definitive study by David Beaudoin and Tim B. Swartz in 2010, which suggested that an even more aggressive strategy, close to the four-minute mark, could be in order for lower-scoring teams.[129]

Is all of this research making a dent? Based on the data tracked

---

127. Andrew Thomas of War on Ice built such a tool, but it went off-line when he was hired by the Minnesota Wild. However, it was not automated.

128. Donald G. Morrison, "On the Optimal Time to Pull the Goalie: A Poisson Model Applied to a Common Strategy in Ice Hockey," *Management Science*, Special Issue on Sports (1976), 137–144.

129. David Beaudoin and Tim B. Swartz, "Strategies for Pulling the Goalie in Hockey," http://people.stat .sfu.ca/~tim/papers/goalie.pdf.

— DOC EMRICK PLAY-BY-PLAY BINGO —

by Noah Davis and Michael Lopez, we know that NHL teams are pulling their goalies earlier than they once did, from an average of under a minute for most of the league's history to 1:18 in the 2014–15 season.[130] However, that's nowhere near the optimal strategy.

Obviously, there may be a reluctance to pull the goalie too soon in the NHL. By going against conventional wisdom, a coach could lose his job and demoralize his team if he gambles and fails too often. Heck, even in my own rec league tournament, I tried to pull our goalie with three minutes left, and he wouldn't budge until there was a minute left. Even then, the sixth player was more concerned about blocking an empty-net attempt than interested in actually helping the team score.

There may be a huge mindset hurdle to overcome, but pulling the goalie earlier makes perfect sense when you think about it. If a team pulls the goalie with three minutes left and the opposing team scores an empty-net goal, then the team's chance of winning the game went from about 10% to almost 0%. However, if they score a goal at 6-on-5, then their chance of winning goes from 10% to 50%. In short, if you make the right call one time out of six, you're breaking even. In time, coaches will start giving this a try, and we'll see for sure who is right and wrong.

130. Noah Davis and Michael Lopez, "NHL Coaches Are Pulling Goalies Earlier than Ever," Five Thirty Eight, October 8, 2015, https://fivethirtyeight.com/features/nhl-coaches-are-pulling-goalies-earlier-than-ever/.

With 269 goals in his first five seasons, one of the most common questions we received in the early days of *Hockey Prospectus* was if Alex Ovechkin could catch Wayne Gretzky's career total of 894 goals.[131] Interest faded when Ovechkin scored 70 goals over the next two seasons and the prospect appeared less likely, but it flared up again when he led the NHL in goals for four straight seasons and had accumulated 525 before his 31st birthday. So, can he reach Gretzky?

Scoring 369 goals past his age of 31 would be an incredible achievement, especially since he scored just 33 goals in 2016–17, but it's not unprecedented. It's probably unsurprising that the record is held by Gordie Howe, who played until he was 51 years old and scored 383 goals in 921 games past age 31. Playing that many games keeps Ovechkin playing in the NHL well into his 40s, which Jaromir Jagr has proven is possible. That makes this a legitimate question worthy of closer study.

We need two things to project Ovechkin's future scoring: his current scoring rate and an age curve. The former is a little tricky, since it fell to 0.40 goals per game in 2016–17 after four consistent seasons between 0.63 and 0.67. I've decided to use his career average of 0.606 goals per game, since it doesn't completely discount his poor season, but it isn't far removed from that optimistic established level.

As described in the opening chapter in *Stat Shot*, age curves are built by forming matched pairs of seasons and calculating the average change at each age, in terms of a percentage.[132] That is, we identify every forward who played in a certain number of games at age 31 and age 32 and calculate the average change in their goals per game. Once we figure out the percentage drop, we apply that to Ovechkin's goal-scoring rate of 0.606 goals per game. Repeat that for every age, and we can chart his future.

Looking at all forwards since the 1967 expansion, a player's goals

131. Rob Vollman, "Howe and Why: Will Ovechkin Outscore Gretzky?," *Hockey Prospectus* (blog), March 11, 2010, http://www.hockeyprospectus.com/puck/article.php?articleid=488.

132. Rob Vollman, "What's the Best Way to Build a Team?," *Stat Shot*, 32–36.

per game increasds until age 23, held that peak until age 25, decreased by around 6% until age 29, and then by around 10% to age 32, after which point it became prone to a collapse of around 15% per season.

Applying this age curve to Ovechkin leaves him well short of the target. At age 38, his goal-scoring will be down to 19 goals per season, and he'll have 776 career goals, which is still 118 goals short of the target. Unless he can play six more seasons as a 20-goal depth-line player, he won't catch Gretzky.

Even if we throw away 2016–17 and arbitrarily assume the best-case scenario—that his actual current goal-scoring rate is his average of 0.65 goals per game from 2012–13 through 2015–16—he still falls to 20 goals per season at age 38 and reaches 792 goals, which is still 102 goals away from the target.

Of course, this age curve is based on the performance of an average NHL player, and there's no way to predict how each individual player will perform. For example, the legendary Johnny Bucyk scored 227 goals in his first 755 games to age 31 and then 329 goals in 785 games through age 42. If Ovechkin is one of those rare talents who can defy Father Time and continue to score up to 40 goals per season through his entire 30s, then Gretzky's record is within reach. And if the NHL institutes new rules that significantly increases scoring, then all bets are off (which is a nice segue to the next question).

Those caveats aside, the statistically most probable result is that Ovechkin will score another 200 goals or more but still finish about 100 goals short of Gretzky's total.

How Can the NHL Boost Scoring?

With the number of goals per game dropping from around 8.0 goals per game in the 1980s to a modern-day peak of 6.2 goals per game in 2005–06 and all the way down to 5.5 in 2016–17, it's no wonder that exploring ways to increase scoring is such a hot topic right now.

I can definitely understand the desire for more scoring, which could make for more exciting games and reduce the reliance on coin tosses like 3-on-3 overtime and a shootout to determine a winner, and it makes it harder for the better team to lose because of one or two bad breaks. But which rules will have the desired effect?

The most common suggestion is to restrict the size of goalie equipment, which is something the NHL has tinkered with for years. In fact, there have been significant regulation changes in four of the past seasons, in addition to allowing more curve to the shooter's sticks, but without any discernible impact on league scoring levels.

Proponents of smaller goaltending equipment point to the high-scoring 1980s, when goalie pads were limited to a width of just 10 inches (25 cm) and the league-average save percentage was a lowly 0.880. Within five years of pads being widened to 12 inches (30 cm) in 1989, save percentages were up over 0.900, and goal-scoring dropped by about 25%. But were the larger pads really to blame for the so-called dead puck era? No. A closer look at the numbers reveals that this decline in scoring had far more to do with improved goaltending technique and defensive play, according to goaltending statistician Phil Myrland.[133] At best, restricting the size of the pads could create an approximate increase of 400 goals a season league-wide, or 0.3 goals per game, and increase the risk of injury.

Instead of shrinking the size of the gear, some people argue about increasing the size of the net. However, this is completely uncharted territory. Since nets are the same size in every league around the world, and always have been, there is no data on which to base a prediction. Normally, we could look to other sports in situations like these, but there really isn't *any* sport that has gone this route. Other than brief experiments, like the NFL reducing the size of its goal posts for the 2015 Pro Bowl, no sport has actually tried to vary its net size for any significant length of time. There's just no way to predict how much scoring would increase for every inch the nets are widened or heightened.

Taking a closer look at that modern-day peak of 6.2 goals per game in the 2005–06 season, some pundits argue that we could

---

133. Phil Myrland, "Why Goalie Equipment Was Not Responsible for the Scoring Drop," *Brodeur Is a Fraud* (blog), November 15, 2009, http://brodeurisafraud.blogspot.ca/2009/11/why-goalie-equipment-was-not.html.

increase scoring with tighter officiating. After all, the bulk of the extra scoring was with the man advantage, and the greater threat of taking more penalties could theoretically open up the game at even strength as well. There was an average of 11.7 power plays per game in 2005–06, which immediately dropped back down to 9.7 the following season and all the way down to 6.0 in 2016–17. At an average power-play conversion rate of 19%, that's a loss of 1.1 goals per game, or at least 1,300 over an entire season. That's why Neil Greenberg of the *Washington Post* described tighter officiating as a "simple fix" to increase scoring levels.[134]

To increase scoring even further, penalties could be served in their entirety, as they were prior to the 1956–57 season. However, that rule change was less about changing scoring levels and more about increasing league parity by clipping Montreal's wings.

What about the recent faceoff changes? Prior to the 2015–16 season, the home team enjoyed the advantage of placing their stick down last for faceoffs. It is now the offensive team who gets that edge. In theory, that should result in the offensive team winning more faceoffs, setting up more scoring opportunities, and scoring more goals.

In reality, the numbers don't justify the importance placed on faceoffs. While winning an offensive zone draw can set a scoring opportunity in motion, it is just one of several plays that must be successfully executed in order to score. That's why the new faceoff rules haven't made much of a difference at all, according to Arik Parnass's preliminary study.[135] Extrapolating his results, these new rules add just 20 goals over an entire season, which is roughly the same impact as allowing players to redirect shots with their skates.

If you're looking for a simple rule change to increase scoring, switching ends may be the way to go. In the first and third periods,

---

134. Neil Greenberg, "There's a Simple Fix to Increase NHL Scoring and It Has Nothing to Do with the Size of the Goal," *Washington Post*, November 11, 2015, https://www.washingtonpost.com/news/fancy-stats/wp/2015/11/11/theres-a-real-simple-fix-to-increasing-nhl-scoring-and-it-has-nothing-to-do-with-the-size-of-the-goal/.

135. Arik Parnass, "Has the NHL's New Faceoff Rule Increased Goal Scoring?," *Hockey Graphs* (blog), November 11, 2015, http://hockey-graphs.com/2015/11/11/has-the-nhls-new-faceoff-rule-increased-goal-scoring/.

each team's bench is on the defensive side of the ice. In the second period, when the bench is located near the offensive zone, teams have to get the puck in a little deeper to make a line change. Consequently, the average number of shots per 60 minutes rises from 28.0 to 30.6 in the second period, shooting percentages increase from 7.5% to 8.1%, and goal-scoring rates go up from 2.11 goals per 60 minutes to 2.48 (all at even strength). Add it all up, and playing the entire game with the long change could increase scoring by 0.6 goals per game, according to an analyst at Pension Plan Puppets.[136]

We could further increase the impact of playing with the long change by adopting stricter icing rules, like disallowing mid-ice tip-ins as a way to wave off an icing or by forcing short-handed teams to leave their own zone before they can ice the puck, as was tried in the old World Hockey Association in the 1970s.

In a similar vein, the league could increase scoring by making the schedule more onerous. According to Mark Pryor's calculations, assigning every team a dozen three-game road trips and one five-game road trip could increase scoring by 0.16 goals per game, or about 200 goals over the season.[137]

Getting rid of points for an overtime or shootout loss could help even further, since it removes a coach's incentive to adopt a lower-scoring mindset and play for the tie. Right now, the ultimate defensive team would end every game tied and then win roughly half of the games in overtime and/or the shootout. Their fans might fall asleep, but they would lead the league with 123 points.

At this point, maybe we're starting to dig through the weeds. Perhaps the answer is to institute some combination of all of these ideas, both big and small. If so, scoring could increase by over 2,500 goals over the entire season, or around two goals per game. That means that scoring levels could eventually rival the high-scoring 1980s once again.

136. Drag Like Pull, "One Simple Rule Change to Boost NHL Scoring," *Pension Plan Puppets* (blog), November 9, 2015, http://www.pensionplanpuppets.com/2015/11/9/9693092/one-simple-rule-change-to-boost-nhl-scoring-long-change.

137. Matt Pryor, "Can the NHL Schedule a Scoring Increase?," *The Hockey Writers*, September 1, 2016, https://thehockeywriters.com/can-the-nhl-schedule-a-scoring-increase/.

However, there's no guarantee that more scoring will actually result in more exciting hockey. Additional goalie injuries, endless power plays, and the gradual erosion of league parity might actually do more harm than good in the eyes of the fans. Today's NHL may already be the most competitive and exciting organized hockey in history, and there may be other ways to improve the sport than focusing on a few extra goals.

What's the Key Stat for Individual Players?

At the team level, there's no argument that wins and losses are the key stats, followed by goal and/or shot-attempt differential. But what about for players? We don't judge players by the team's win-loss record in games they played (except for goalies, ridiculously enough), and the equivalent to goal differential is the plus/minus statistic, which is rightfully regarded with great skepticism

How about Corsi? Well, Corsi is the equivalent of shot-attempt differential, but it is viewed with skepticism at the individual level for the exact same reason that plus/minus is. Yes, the larger volume of data reduces the impact random variation can have on it, but it is just as contextual as plus/minus. As we explored in the "How Can Stats Be Placed in Context?" chapter, a player on a weak team with weak linemates who faces the top opponents in the defensive zone will necessarily have a worse Corsi than an inferior player on a strong team with strong linemates who faces secondary opponents in the offensive zone. One could argue that a version of Corsi that is adjusted for all of these factors would be the best statistic, but I don't think something that complicated should be the one key stat, especially since it would be incalculable for any league and era beyond today's NHL. For the same reason, I would dismiss expected-goals models and catch-all statistics like GVT and WAR as the key individual player stat.

I think that the answer is ice time. That's the first thing I look at when studying a particular player. You can learn so much from how much ice time a player is assigned, how it breaks down by manpower situation, and where that ranks on the team. Even some of the most informative new metrics can involve variations of ice time in some fashion or another. For example, the amount of ice time a player is assigned in the defensive zone or against top opponents or when the team is protecting a lead can all provide valuable information about a player's usage.

Even Iain Fyffe's point allocation, which was the field's first catch-all stat, estimated a player's defensive abilities by comparing a player's total ice time to how much of it could be explained by how many goals he created. For example, on the 2017 Stanley Cup–champion Pittsburgh Penguins, defenceman Justin Schultz outscored Brian Dumoulin 51 to 15 in the regular season and 13 to 6 in the playoffs, but Dumoulin averaged 19:03 even-strength minutes per game in the post-season and Schultz averaged 16:16. Clearly, the only reason Dumoulin would get that much more ice time is if his defensive contributions were even more considerable than Schultz's scoring.

However, ice time isn't a perfect statistic, since it depends on the team's other options. For example, a highly gifted centre would get a lot more ice time on the Arizona Coyotes than on the Pittsburgh Penguins, who already have Sidney Crosby and Evgeni Malkin. It also depends on the coach's assessment of a player's abilities, which may not always be accurate.

There is no single stat with which to perfectly evaluate an individual player, but ice time is probably the most informative, and it is certainly the best place to start a player evaluation.

SUPER GLOSSARY

It used to be easy to stay on top of hockey's various stats. Because the field was so small, new developments came along very gradually, and there were only a handful of websites you had to visit, or people you had to consult, before you found whatever information you required.

With the growing popularity of hockey analytics over the past decade, it's no longer possible to keep track of all the different metrics, including what they are, how they're calculated, and where they came from. This problem gets worse with every passing season, as new stats are innovated and the online history gradually disappears as key websites are shut down. As a result, stats are constantly misused, go uncredited, or are reinvented.

Ideally, there would be a glossary on every website where these stats are innovated, like *Hockey Graphs*, or where they are hosted, like *Corsica Hockey*, but there's usually no such luck. While everyone agrees that there would be value in publishing a detailed glossary, I've been told it's simply too much effort to accurately describe these stats and trace and record their origins.

Even the NHL, with its big budget and extensive resources, launched its enhanced stats platform in February 2015 without any explanation of where the stats came from or what they mean.[138] In fact, the NHL even renamed some popular stats like Corsi, Fenwick,

---

138. Full disclosure: I have written for NHL.com since the 2015–16 season and have engaged with their enhanced stats team both prior, during, and since the launch.

and PDO, which foiled any ambitious fans' attempts to search for answers elsewhere.

There may also be an incentive issue at play. After making such a big investment, why should the NHL promote the contributions of unaffiliated hobbyists? And if you built a stats website of your own, would you invest your time in adding features to help build your own name and reputation, or in putting together a glossary that builds someone else's?

Beyond the difficulties and conflicts in tracing the meanings and complicated histories of these innovations, there's also the perception that some of our stats have become so mainstream that they no longer require any explanation and are owned by no one. After all, you won't find plus/minus in a glossary, so why include Corsi?

In my view, we *should* credit our sources for stats like plus/minus. Like us, Allan Roth was initially a hobbyist who made important contributions to our sport in his spare time, but how many fans have ever heard of him (or of *any* hockey statistician)? And how many fans actually understand the plus/minus statistic? Quiz yourself right now. You know that it's a player's goal differential, but which types of goals are included? Are empty-net goals included? How about short-handed goals or power-play goals? How about penalty shots? Unless you believe most fans can answer those questions with confidence, maybe plus/minus *should* be included in a glossary. Well, now it is.

Being one of the few people who has been around long enough to acquire the necessary breadth of experience and network of contacts, and being someone who wants to promote the entire field and everyone in it, I eventually landed on the idea of building a comprehensive glossary myself. In fact, with the hundreds of footnotes in previous editions of *Hockey Abstract*, I had already done a lot of the work.

That's why this glossary goes beyond the simple list of statistics and brief explanations included in previous editions of *Hockey Abstract*. It includes every statistic's origin, definition, exact formula (when feasible), and links to help launch quests for more detailed information.

I restricted the scope to the statistics themselves and a few statistical concepts. This glossary is not meant to cover prominent statisticians, websites, technologies, or tools. Similarly, it does not include generic, non-hockey statistical concepts like correlation, confidence intervals, regression, sample size, types of bias, and so on, because those are not specific to our field and are well-documented elsewhere. Finally, there are no generic hockey concepts, like power play, offside, or icing. This glossary is exclusively about the world of hockey analytics.

Even with that specific scope, it still runs quite a few pages, so I emphasized concision, especially with less common or useful stats. In fact, quite a few of these stats are inactive or defunct, while others are admittedly obscure. However, I didn't want to be a gatekeeper of which stats are worthy of inclusion, so I left nothing out. Plus, you never know which ones will get revived or get incorporated into better and more lasting metrics in the future. If nothing else, they're all interesting as a matter of historical record.

Can't find what you're looking for? Some statistical concepts go by several different names, or they overlap and/or lead to others, so look around to see if they are listed elsewhere. Furthermore, I broke down compound statistics into their component pieces. For example, instead of including Fenwick percentage, Fenwick save percentage, on-ice Fenwick, and all of the Corsi counterparts as separate entries, I just listed the component pieces, which are Fenwick, Corsi, percentage stats, save percentage, and on-ice stats.

If you still can't find what you're looking for, then it's quite likely that I have missed a few, for which I apologize. This may be especially true about the more recent statistics, since it's almost impossible to stay on top of every development (and because book manuscripts are written up to a year before they actually hit the shelves).

Despite all the work involved, compiling this glossary was very rewarding and a lot of fun. If you walk away with nothing else, consider how many stats there are, how far back they go, and how many people are involved. Our world is so much bigger than Corsi and PDO, and it's getting bigger every day. Enjoy!

3-1-1 RULE

The rule of thumb is that roughly every three goals that a team scores or prevents results in one additional point in the standings and costs about one million dollars in cap space (and rising). The earliest formal citing of this rule was by me in 2012, but it was built on concepts that several people had independently established at least three years prior.[139]

60-40 POSSESSION RULE

First formally advanced by Ben Wendorf in January 2014, the 60-40 rule states that almost every player and team will have Fenwick percentages that fall between 40% and 60% over the long term, regardless of any outside factors.[140]

ABOVE-AVERAGE APPEARANCE PERCENTAGE (AAA%)

In April 2016, Nick Mercadante introduced a quality starts variation that takes shot quality into account.[141] Goalies are awarded an above-average appearance if they stop at least as many shots in a single game as a league-average goalie who faced the same number and quality of shots.

ADJUSTED CORSI

While there are several adjusted forms of Corsi, this term is normally reserved for Michael Parkatti's version from March 2013.[142] Parkatti took a player's existing zone-adjusted Corsi and made the first-known attempt to further adjust it for the quality of his linemates and his opponents. The formula is as follows: actual Corsi per 60 minutes plus 11.91 plus quality of competition minus quality of teammates minus 0.24 times his zone start percentage equals adjusted Corsi.

ADJUSTED PLUS/MINUS (APM)

Plus/minus has its flaws, and fixing them was a rite of passage in

139. Rob Vollman, "Goals Versus Salary, 2011–12," *Hockey Prospectus* (blog), October 10, 2012, http://www.hockeyprospectus.com/puck/article.php?articleid=1393.

140. Ben Wendorf, "A Rule of 60-40: Thoughts on Individual Player Possession Metrics," *Hockey Graphs* (blog), January 22, 2014, https://hockey-graphs.com/2014/01/22/nhl-possession-rule-of-60-40-fenwick-corsi/.

141. Nick Mercadante (@NMercad), "High Above Average Apperance % ("AAA%") correlates to High Win Threshold %. r2=.47. So look for consistency first," Twitter, April 9, 2016, 5:12 p.m., https://twitter.com/nmercad/status/718954674406055936.

142. Michael Parkatti, "Adjusting Corsi for Zone Starts and Quality Factors," *Boys on the Bus* (blog), http://www.boysonthebus.com/2013/03/14/adjusting-corsi-for-zone-starts-and-quality-factors/ (site discontinued).

the early days of hockey analytics. The simplest form of adjusted plus/minus is restricted to even-strength play and removes the team average from each player, as first introduced by Klein and Reif in 1986 and further refined by Iain Fyffe and me in 2001.[143] The more sophisticated versions are measured on a per-minute basis and adjusted for a player's linemates and opponents, as introduced by Timo Seppa in August 2009 as ESTR and adapted by Brian Macdonald in 2011, who used a regression-based model.[144] Versions that don't use goal-based data—and instead use shot quality data to calculate weighted shots or expected goals, like Cam Lawrence introduced in 2014—are now more commonly classified as expected plus/minus.[145]

ADJUSTED SAVE PERCENTAGE (ADJSV%)

What would a goalie's save percentage be if he faced a typical distribution of shots? While there are many adjusted forms of save percentage, this term usually refers to the quality-adjusted version introduced by Andrew Thomas and Sam Ventura in 2014 as "the weighting of a goaltender's save percentage in each danger level by the fraction of shots that would be expected from the league-wide distribution."[146]

ADJUSTED STATISTICS (ADJ PREFIX)

Some individual and team statistics are adjusted for outside factors, such as manpower situation, team strength, score effects, shot quality, zone starts, quality of linemates and/or opponents, era, and so on. Statistics are usually named after what they're adjusted for, like score-adjusted Corsi, but others are adjusted for too many factors to be so named.

143. Jeff Z. Klein and Karl-Eric Reif, "Plus/Minus: The Red and the Black," *The Klein and Reif Hockey Compendium* (Toronto: McClelland and Stewart, 1986) 96–130; Iain Fyffe with Rob Vollman, "Improving Plus-Minus Making Statistics More Useful," *The Hockey Research Journal* V., no. 1 (2001): 38–42.

144. Timo Seppa, "Driving to the Net, Even Strength Total Rating," *Hockey Prospectus* (blog), August 31, 2009, http://www.hockeyprospectus.com/puck/article.php?articleid=254; Brian Macdonald, "A Regression-Based Adjusted Plus-Minus Statistic for NHL Players," *Journal of Quantitative Analysis in Sports* 7, Issue 3, Article 4 (2011), http://hockeyanalytics.com/Research_files/Regression_Based_Plus_Minus.pdf.

145. Cam Lawrence, "Adjusted Plus-Minus: Measuring Defensive Effectiveness," *Canucks Army* (blog), October 20, 2014, http://canucksarmy.com/2014/10/20/adjusted-plus-minus-measuring-defensive-effectiveness.

146. Greg Balloch, "An Introduction to Adjusted Save Percentage," *InGoal Magazine*, September 7, 2015, http://ingoalmag.com/analysis/an-introduction-to-adjusted-save-percentage/; Andrew Thomas, "Annotated Glossary," *WAR on Ice* (blog), November 26, 2015, http://blog.war-on-ice.com/annotated-glossary/.

AGING CURVE

When projecting a player's performance, aging curves are used to adjust for the expected rise and decline with age. The most accurate method is the matched pair system popularized in 2003 by baseball statisticians, and which I explained in more detail in *Stat Shot*.[147]

ALL THREE ZONES PROJECT

In April 2014, Corey Sznajder launched the All Three Zones Project, which involved manually "tracking zone entries and exits for every game of the 2013–14 NHL season."[148] The end result is a data set that has improved our understanding of what drives puck possession.

ASSIST (A)

Assists, which were first officially recorded by the NHL during the 1918–19 season, have undergone slight changes over the past century, and they have always involved at least an element of subjectivity and/or scorekeeper bias. Today, the most noteworthy distinction to make is between the primary assist and the secondary assist. If the data is ever made available, it would also be interesting to break down the primary assists on goals that were scored from direct passes, passes and carries, deflected shots, and rebounds.

AYNAY (AGAINST YOU OR NOT AGAINST YOU)

AYNAY is essentially the same stat as WOWY (see page 304), but for opposition instead of linemates.

BAYES-ADJUSTED FENWICK CLOSE

Lacking enough data on which to evaluate teams early in the season using the Fenwick close statistic, in October 2014, a blogger going by the pseudonym garik16 introduced a version that includes weighted data from the previous season.[149]

BLOCKED SHOT (BS OR BKS)

Introduced for the 1997–98 season with the rest of the NHL's new RTSS statistics and officially recorded since 2002–03, a blocked shot

147. Tom Tango, "Forecasting Pitchers – Adjacent Seasons," *Tango Tiger* (blog), 2003, http://www.tangotiger.net/adjacentPitching.html; Vollman, "What's the Best Way to Build a Team," *Stat Shot*, 32–36.

148. Corey Sznajder, "All Three Zones Tracking Project," GoFundMe, April 15, 2014, https://www.gofundme.com/allthreezones.

149. garik16, "Bayes-Adjusted Fenwick Close Numbers – an Introduction," *Hockey Graphs* (blog), October 18, 2014, https://hockey-graphs.com/2014/10/18/bayes-adjusted-fenwick-close-numbers-an-introduction/.

is awarded whenever a defending player prevents a shot from reaching the net. This is a raw count and is generally adjusted for some combination of manpower situation, scorekeeper bias, and the number of shot attempts faced before being used to evaluate players or teams.

BOX CAR

A common abbreviation of a newspaper's box score stats, a player's box cars generally refer to traditional player stats like games played, goals, assists, points, plus/minus, and penalty minutes, as well as goalie stats like wins and losses, goals-against average, save percentage, and shutouts.

BOX PLUS/MINUS (BPM)

Based on the basketball statistic of the same name developed by Daniel Myers in October 2014, Dawson Sprigings introduced BPM to hockey in October 2016 as a way to measure a player's performance relative to his replacement level using a weighted sum of the information contained in modern NHL box scores.[150]

BRADLEY-TERRY MODEL

This is a chess rating system innovated in the 1920s by Ernst Zermelo and then reinvented independently by R.A Bradley and M.E. Terry in 1952 for taste tests. It can be applied to any pairwise model, and it formed the basis for the KRACH model Ken Butler applied to U.S. college hockey in the 1990s.

CAP EFFICIENCY CHART

One of the total performance charts I introduced for *Hockey Prospectus 2013–14*, a team's cap efficiency chart maps each player's annual cap hit on the horizontal axis against his GVT performance on the vertical axis with circles sized and shaded based on the resulting GVS.[151] More recently, stats like WAR are used on the vertical axis.

CAP HIT OF INJURED PLAYERS (CHIP)

Introduced by Thomas Crawshaw in April 2009, CHIP estimates the total impact of a team's injuries by multiplying the man-games lost by the game-based cap hit of the players who were injured.[152]

---

150. Daniel Myers, "About Box Plus/Minus (BPM)," Basketball Reference, October 2014, http://www.basketball-reference.com/about/bpm.html; Dawson Sprigings, "Introducing Box Plus-Minus," *Hockey Graphs* (blog), October 26, 2016, https://hockey-graphs.com/2016/10/26/introducing-box-plus-minus/.

151. Rob Vollman, "Introducing Our Total Performance Charts," *Hockey Prospectus 2013–14* (2013), x–xv.

152. Thomas Crawshaw, "Ow, That Really Hurts!," *Springing Malik* (blog), April 20, 2009, http://

CATCH-ALL STATISTICS

Any statistic that attempts to capture all of a player's value in a single metric. See also point allocation (page 283), GVT (page 273), player contribution (page 281), point shares (page 283), total hockey rating (page 300), expected goals (page 267), and WAR (page 304).

CLOCK-ADJUSTED STATISTICS

Since score effects become more significant as a game wears on, analysts like Micah Blake McCurdy (in November 2014[153]) have tinkered with the idea of refining the score adjustments used with shot-based metrics based on the time remaining on the clock.

CLOSE GAME (OR CLOSE SCORE) STATISTICS

A concept that dates at least as far back as 2009, when proposed by Gabriel Desjardins, close game posits that stasticians can avoid the skewing effect of how teams play for the score by considering only close-game situations, which was defined by Tore Purdy in October 2010 as "whenever the score is within one goal in the first or second period, or tied in the third period, or overtime."[154] It has since been overtaken by score-adjusted stats in terms of accuracy, but it's still the preferred method for simplicity.

CORDELIA

Micah Blake McCurdy unveiled the Cordelia single-game prediction model in September 2016 as a replacement for Oscar, his previous model, which replaced Pip, his first model. Using 19 inputs, "Cordelia is a logistic regression model that estimates the probability with which the home team will win the game based on recent measurements of the skill of the two teams.[155]

---

springingmalik.blogspot.com/2009/04/ow-that-really-hurts.html.

153. Micah Blake McCurdy, "Adjusted Possession Measures," *Hockey Graphs* (blog), November 13, 2014, https://hockey-graphs.com/2014/11/13/adjusted-possession-measures/.

154. Tore Purdy, "Corsi Corrected for Schedule Difficulty," *Objective NHL* (blog), October 23, 2010, http://objectivenhl.blogspot.ca/2010/10/corsi-corrected-for-schedule-difficulty.html/.

155. Micah Blake McCurdy, "Prediction Model Cordelia," HockeyViz, September 25, 2016, http://hockeyviz.com/txt/cordelia; Micah Blake McCurdy, "Prediction Model Oscar," HockeyViz, October 6, 2015, http://hockeyviz.com/txt/oscar; Micah Blake McCurdy, "Season Simulations with Pip," *Silver Seven Sens* (blog), December 23, 2014, http://www.silversevensens.com/2014/12/23/7439287/season-simulations-with-pip.

CORE AGE

Core age is the average age of a team's players weighted by a given statistic. In the case of Timo Seppa's September 2011 incarnation of core age, each player's age is multiplied by his GVT and then the sum of all players' results is divided by the team's GVT.[156]

CORSI

Also known as SAT and historically as net shots, Corsi in its purest form is simply a count of all shot attempts, including those that went in, were saved, were blocked, or missed the net. It is sometimes used as a synonym for shot attempts, and it is sometimes meant as a differential between a team's shot attempts minus those of its opponents. It can also be expressed as a percentage-based stat or as a rate-based stat. Given its popularity and usefulness, there are a number of adjusted and relative forms of this statistic. Commonly believed to have been invented by former Buffalo goalie coach Jim Corsi sometime between 2001 and 2009, its roots actually go back at least three decades earlier; it was named and introduced to the mainstream by Timothy Barnes.[157]

COUNTING STATISTICS

In the world of statistics, count data is based on an accumulation of observed events that either occurred in full or did not occur at all. In the world of hockey, good examples include shots, goals, and wins.

DANGEROUS FENWICK (DFF)

See expected goals (page 267) and weighted shots (page 303).

DANGEROUS PRIMARY SHOT CONTRIBUTION (DPSC)

As defined by Alan Wells in August 2016, DPSCs are passes that either cross the royal road or come up from behind the net and are immediately followed by a shot attempt. [158]

DEFENCE-INDEPENDENT GOALIE RATING (DIGR)

Inspired by baseball's defence-independent pitching statistics (DIPS), developed by Voros McCracken in 1999, Michael Schuckers

---

156. Timo Seppa, "Core Age and the Strategic Direction of Teams," *Hockey Prospectus 2011–12* (2011), 404–410.

157. Bob McKenzie, "McKenzie: The Real Story of How Corsi Got Its Name," TSN, October 6, 2014, http://www.tsn.ca/mckenzie-the-real-story-of-how-corsi-got-its-name-1.100011.

158. Alan Wells, "Passing Project: Dangerous Primary Shot Contributions," *NHL Numbers* (blog), August 10, 2016, http://nhlnumbers.com/2016/8/10/passing-project-dangerous-primary-shot-contributions.

introduced DIGR to hockey in March 2011 as a method of calculating a goalie's performance independently of the quality of shots allowed by the team.[159] This location-adjusted version of save percentage breaks down a goalie's performance by shot location and demonstrates what the save percentage would be if he had faced the league-average distribution of shots instead.

DEFLECTION AND TIP

Since the 2002–03 season, the NHL has recorded the shot types in their game files, which include whether or not the official scorekeeper felt the shot was intentionally tipped or inadvertently deflected.

DELTA

Sometimes referred to as deltaSOT because it is adjusted for situation, opponents, and teammates,[160] delta is the first of several shot-based context-adjusted statistics. It was developed by Tom Awad in January 2010.[161]

DELTA CORSI (DCORSI)

The spiritual successor to delta is Stephen Burtch's dCorsi, which he unveiled in July 2014.[162] The calculation involves establishing how many shot attempts for and against are expected for each player. It is based on a linear regression of several variables, such as the player's position, zone starts, linemates, and opponents, and then compares that against the actual results.

DEV

Inspired by PCS, the DEV model is a historical player projection system that uses similarity scores. Introduced by Zac Urback and Hayden Speak in March 2016, DEV estimates a prospect's chances of

159. Michael Schuckers, "DIGR: A Defence Independent Rating of NHL Goaltenders Using Spatially Smoothed Save Percentage Maps" (paper presented at MIT Sloan Sports Analytics Conference, March 4, 2011), http://www.hockeyanalytics.com/Research_files/DIGR_Schuckers.pdf.

160. Tom Awad, "Numbers on Ice, Delta with Teammate Adjustments – DeltaSOT," *Hockey Prospectus* (blog), February 5, 2010, http://www.hockeyprospectus.com/puck/article.php?articleid=454.

161. Tom Awad, "Numbers on Ice, Plus-Minus and Corsi Have a Baby," *Hockey Prospectus* (blog), January 21, 2010, http://www.hockeyprospectus.com/puck/article.php?articleid=436.

162. Steve Burtch, "dCorsi–Introductions," *NHL Numbers* (blog), July 19, 2014, http://nhlnumbers.com/2014/7/19/dcorsi-introductions.

success by combing CHL history for players of similar size with similar era-adjusted statistics at the same age, to see how well they performed.[163]

THE DEVASTATION SCALE

As an alternative way to identify the best teams in history, Jeff Z. Klein and Karl-Eric Reif introduced the devastation scale in their 1986 *Compendium* as a team's goals scored divided by goals against.[164] The devastated scale is the same, but goals against are divided by goals scored. These days, goal percentages are used instead.

DISCIPLINED AGGRESSION PROXY (DAP)

In October 2001, Iain Fyffe introduced the disciplined aggression proxy, which is calculated by taking a player's aggression, as measured by adding hits and takeaways, and dividing it by discipline, as measured by penalty minutes or unmatched minor penalties.[165]

DO-IT-ALL INDEX

Introduced by me in May 2011, the do-it-all index awards a single point for fulfilling each of 10 statistical criteria based on different types of individual player contributions, like scoring, faceoffs, special teams, and shootout.[166]

DRAFT PICK VALUE CHART

In the early 1990s, Dallas Cowboys coach Jimmy Johnson assigned a value to each position in the NFL draft. Dozens of these draft pick value charts have since been created for hockey, based on the past success of players at the various draft positions. To my knowledge, hockey's first public attempt was by TSN's Scott Cullen in February 2009 and, privately, around the same time by Josh Flynn of the Columbus Blue Jackets.[167]

---

163. Zac Urback, "Introducing DEV," *Prospect Stats* (blog), March 2016, http://prospect-stats.com/blog/Introducing_DEV/.

164. Jeff Z. Klein and Karl-Eric Reif, "The Best Teams," *The Klein and Reif Hockey Compendium*, 47–48.

165. Iain Fyffe, "Puckerings Archive: Search for Meaning in RTSS (22 Oct. 2001)," *Hockey Historysis* (blog), August 29, 2014, http://hockeyhistorysis.blogspot.ca/2014/08/puckerings-archive-search-for-meaning.html.

166. Rob Vollman, "Howe and Why: Do-It-All Players," *Hockey Prospectus* (blog), May 19, 2011, http://www.hockeyprospectus.com/puck/article.php?articleid=949.

167. Scott Cullen, "Backchecking: The Value of NHL Draft Picks," TSN, February 20, 2009, http://www.tsn.ca/columnists/scott_cullen/?ID=267960 (page no longer available); Rob Mixer, Analytics, "Value Chart" Bring New Dynamic to Jackets' Draft Day Strategy," National Hockey League, June 17, 2014, https://www.nhl.com/bluejackets/news/analytics-value-chart-bring-new-dynamic-to-jackets-draft-day-strategy/c-722857.

EFFICIENCY

See save percentage (page 290).

ELO RATING SYSTEM

Originally invented by Arpad Elo in the late 1950s to replace the Harkness system as a way of rating chess players, the Elo rating system has since been applied to a number of different types of head-to-head competitions. In hockey, Elo variations have been used to rate teams in college hockey, international competitions, and the NHL, and it has also been used as a crowd-sourced way to rank players.

EMPTY-NET GOAL (ENG)

Any goal scored while a team is playing without its goalie—which is usually for a delayed penalty or for an extra attacker late in the game—is termed an empty-net goal. The first recorded instance of pulling the goalie was by Boston's Art Ross on March 26, 1931, after which time scorekeepers informally and inconsistently found ways to note ENGs in order to keep goaltending data accurate. However, ENGs are still included in a skater's data, including in shooting percentage calculations and when they were simply the last person to touch the puck before a defensive player's own goal.

ERA-ADJUSTED STATISTICS

Since scoring levels have changed over the years, from a high of 9.58 in 1919–20 to a low of 2.92 goals per game in 1928–29, scoring statistics should be adjusted before data from different seasons are compared. The most popular method, as introduced by Dan Diamond in 1999, is to divide a player's goals and assists by that season's league-average goals per game and then to multiply that total by the current standard.[168] Though not as widely adopted, Diamond's approach also adds the refinement of removing the player's own scoring from the calculation.

EVEN-STRENGTH BLOCKED SHOTS PERCENTAGE (ESBS%)

To account for the greater volume of shot attempts that certain shot-blockers face, Derek Zona introduced ESBS% in December 2011, which is the percentage of shots a player has blocked among all on-ice shot attempts faced at even strength.[169] He credited

---

168. Dan Diamond, "Adjusted Scoring," *Total Hockey*, 626.

169. Derek Zona, "Which Oilers Really Block the Most Shots," *Copper & Blue* (blog), December 14, 2011, http://www.coppernblue.com/2011/12/14/2636304/oilers-block-the-most-shots.

George Ays and Gabriel Desjardins for the idea. In September 2016, Iain Fyffe introduced a version called RESB% that was regressed to account for those who have faced a lower volume of shot attempts.[170]

EVEN-STRENGTH POINTS PERCENTAGE (ESP%)

See individual points percentage (page 275).

EVEN-STRENGTH STATISTICS (ES PREFIX)

Since shooting and scoring go way up when a team has the man advantage, and since some players are assigned a lot of time on special teams while others get none, it makes sense to compare players in even-strength situations only. In fact, shot-based metrics are done almost exclusively this way. While most even-strength stats refer strictly to five-on-five situations, some include three-on-three and/or four-on-four hockey as well as empty-net situations.

EVEN-STRENGTH TOTAL RATING (ESTR)

Introduced by Timo Seppa in August 2009, ESTR is a form of adjusted plus/minus that represents "the number of goals per 60 minutes that the player is worth at even strength, given NHL-average players as teammates and opposition."[171]

EXPECTED GOALS (XG, xG, OR EG)

Following the inclusion of shot location data in NHL game files for the 2002–03 season, in January 2004 Alan Ryder introduced the notion of calculating how many goals are expected based on the number and quality of the shots that were taken.[172] Almost a decade later, a flurry of expected-goals models was introduced, most notably by Brian Macdonald in March 2012, Wesley Yue in August 2013 (as LAEGAP), Matt Cane in March 2014, Matt Pfeffer in June 2014, Peter Tanner in March 2015, Nick Abe in March 2015, Dawson Sprigings in May 2015, GMoney in October 2015 (as Dangerous Fenwick), and Emmanuel Perry in March 2016, with the key differences being how they measured

170. Iain Fyffe, "Who Is the Best Shot-Blocker?," *Stat Shot*, 148–157.

171. Timo Seppa, "Driving to the Net, Even Strength Total Rating," *Hockey Prospectus* (blog), August 31, 2009, http://www.hockeyprospectus.com/puck/article.php?articleid=254.

172. Alan Ryder, "Shot Quality" (PDF), Hockey Analytics, January 2004, http://hockeyanalytics.com /Research_files/Shot_Quality.pdf.

the quality of the shots.[173] Rather than serving as stand-alone statistics, expected goals are usually incorporated into goaltending metrics like DIGR, catch-all statistics like player contribution, adjusted forms of plus/minus, and, most commonly, as variations on shot-based metrics (like Corsi) that include that shot quality component. They are also sometimes referred to as weighted shots.

EXPECTED PLUS/MINUS

To address the impact luck can have on the relatively small number of goals that are scored while a player is on the ice in any given season, a popular application of expected-goals models is in an alternate version of plus/minus. We calculate expected plus/minus using the number of expected goals (or weighted shots) based on the number and quality of shots taken while a player was on the ice, rather than the actual number of goals. Cam Lawrence was among the first to use expected goals this way (in October 2014), and Dawson Sprigings unveiled a version in October 2016, but the currently accepted version was introduced by Mike Lynch in October 2015, based on Wesley Yue's LAEGAP.[174]

---

173. Brian Macdonald, "An Expected Goals Model for Evaluating NHL Teams and Players" (paper presented at MIT Sloan Sports Analytics Conference, March 2, 2012), http://www.hockeyanalytics.com/Research_files/NHL-Expected-Goals-Brian-Macdonald.pdf; Wesley Yue, "LAEGAP Methodology," *Hockey Metrics* (blog), August 28, 2013, http://hockeymetrics.net/laegap-methodology/; Matt Cane, "Introducing xGF20: A Context Neutral, Corsi-Based Goal Creation Metric," *Puck Plus Plus* (blog), March 27, 2014, https://puckplusplus.com/2014/03/27/introducing-xgf20-a-context-neutral-corsi-based-goal-creation-metric/; Matt Pfeffer, "Exploring the Theoretical Limits of Shot Quality," *Hockey Prospectus* (blog), June 4, 2014, http://www.hockeyprospectus.com/exploring-the-theoretical-limits-of-shot-quality/; Peter Tanner, "Playoff Probabilities and Season Simulator," Money Puck, March 2015, http://moneypuck.com/about.htm; Nick Abe, "eGF Finally Fully Explained," *Xtra Hockey Stats* (blog), January 14, 2016, http://xtrahockeystats.com/wordpress/?p=70; Dawson Sprigings, "Updated NHL Expected Goals Model," *Don't Tell Me About Heart* (blog), May 25, 2015, http://donttellmeaboutheart.blogspot.ca/2015/05/updated-nhl-expected-goals-model.html; GMoney, "Explaining Dangerous Fenwick," *Oilers Nerd Alert* (blog), October 30, 2015, https://oilersnerdalert.wordpress.com/2015/10/30/explaining-dangerous-fenwick/; Emmanuel Perry, "Shot Quality and Expected Goals: Part I," *Corsica Hockey* (blog), March 3, 2016, http://www.corsica.hockey/blog/2016/03/03/shot-quality-and-expected-goals-part-i/.

174. Cam Lawrence, "Adjusted Plus-Minus: Measuring Defensive Effectiveness," *Canucks Army* (blog), October 20, 2014, http://canucksarmy.com/2014/10/20/adjusted-plus-minus-measuring-defensive-effectiveness; Dawson Sprigings, "Introducing Expected Plus-Minus," *Hockey Graphs* (blog), October 25, 2016, https://hockey-graphs.com/2016/10/25/expected-plus-minus/; Mike Lynch, "Introducing NHL Expected +/–," Sports Reference, October 7, 2015, http://www.sports-reference.com/blog/2015/10/introducing-nhl-expected/.

EXPECTED SAVE PERCENTAGE (XSV%)

As first pioneered with the SQNSV statistic, a popular application of the expected-goals model is to evaluate goalies by comparing how they actually performed to what was expected, given the quality of shots they faced. In December 2007, Gabriel Desjardins first defined xSV% as how a league-average goalie would have fared against the same quality of shots faced by a particular goalie; it is calculated by dividing expected goals by shots faced.[175] This statistic was later revised and reintroduced by Dawson Sprigings in March 2015.[176]

EXPECTED STATISTICS (EXP OR X PREFIX)

Normally, expected stats refer specifically to those using an expected-goals model. However, sometimes a particular statistic is placed in context by comparing it to how the team or player was expected to perform based on that season's league averages as well as specific to any number of other circumstances, such as position, age, manpower situations, score effects, shot quality factors, linemates, or opponents.

FACEOFF PERCENTAGE (FO%)

One of the few *over*-studied areas in hockey, a player's faceoff percentage is the number of faceoffs won divided by the total number of faceoffs. Officially recorded since 1997–98, faceoff wins are awarded based on the scorekeeper's opinion about which team touched the puck first after a faceoff, other than the centres themselves. As such, it can be biased, doesn't allow for ties, and doesn't necessarily reflect which team actually first gained control of the puck. These faceoff stats can be broken down into specific situations, like zone, manpower situation, handedness, or opening faceoffs only. There are also stats that measure what percentage of a team's draws a given player takes and a team's faceoff percentage when a given player is on the ice.

FENWICK (ALSO USAT)

Fenwick is a popular variation on Corsi that excludes blocked shots. It was named by Timothy Barnes after it was suggested by Matt

---

175. Gabriel Desjardins, "2007–08 5v5 Goaltender Performance," *Behind the Net* (blog), December 23, 2007, http://www.behindthenet.ca/blog/2007/12/2007-08-5v5-goaltender-performance.html.

176. Dawson Sprigings, "xSV% – Save Percentage Accounting for Shot Quality," *Don't Tell Me About Heart* (blog), March 13, 2015, http://donttellmeaboutheart.blogspot.ca/2015/03/xsv-save-percentage-accounting-for-shot.html.

Fenwick in November, 2007. Conceptually, some analysts want to give a player credit for blocking a shot rather than letting it through. Statistically, blocking shots was found to be a persistent skill, and Fenwick was observed to have a closer relationship with scoring-chance data than Corsi.

FENWICK CLOSE

See close game statistics (page 262).

FENWICK SHOOTING AND SAVE PERCENTAGE

See true shooting and save percentage (page 301).

FIRST GOAL (1G)

More of a curiosity than a statistic of any value, a first goal is awarded to the player who scored the first goal of the game, unless it was in a game-deciding shootout. In December 2008, Tore Purdy demonstrated that it is no more important than the second goal.[177]

FISCHLER POINT (FPTS)

Named by Charles Mousseau as a homage to the original idea introduced by Stan and Shirley Fischler, a Fischler Point is a goal or an assist weighted by the score at the time it occurred.[178]

FLEMWICK (ADJGSAXGA/60)

See Mercad (page 277).

GAME SCORE (GMSC)

How well did a player perform in a given game? Originally developed by Bill James in 1986 for baseball as a measure of a pitcher's performance in a single game, game score was introduced to hockey by Dominik Luszczyszyn in July 2016 as a weighted average of 13 statistics found in a single game summary.[179]

GAME STAR

First awarded in the 1936–37 season, first, second, and third game stars

177. Tore Purdy, "The First Goal," *Objective NHL* (blog), December 20, 2008, http://objectivenhl .blogspot.ca/2008/12/first-goal.html.

178. Charles Mousseau, "Who Is the Best Clutch Scorer?," *Hockey Abstract 2017*, 193; Stan and Shirley Fischler, *Breakaway '86: The Hockey Almanac* (1986), 9–10.

179. Dominik Luszczyszyn, "Measuring Single Game Productivity: An Introduction to Game Score," *Hockey Graphs* (blog), July 13, 2016, https://hockey-graphs.com/2016/07/13/measuring-single-game-productivity-an-introduction-to-game-score/.

are the scorekeeper's assessment of the three most outstanding players of the game.

GAME-WINNING GOAL (GWG)

If N is the number of goals scored by the losing team, then the N+1st goal scored by the winning team is considered the game-winning goal, unless it occurred in a shootout.

GIVEAWAY

An RTSS statistic (see page 287) introduced in 1997–98, a giveaway is commonly believed to be a subjective measure of when a player's actions result in the loss of puck possession to the opposing team, but there are far too few of them recorded. Seasoned trackers can easily count 20 turnovers for every giveaway recorded in NHL game files.

GOAL-BASED STATISTICS

Traditionally, statistics that measure team and player performance involve goals, like plus/minus, goals-against average, power-play percentage, and so on. There are relatively few goals per game, but they have an almost perfect relationship with wins and losses and require no subjective opinions to accurately record.

GOALS-AGAINST AVERAGE (GAA)

One of hockey's earliest goaltending statistics, GAA is calculated by dividing goals allowed by minutes played and then multiplying that result by 60 minutes.

GOALS ALLOWED %- (GA%-)

See save percentage+ (page 290).

GOALS CREATED (GC)

A player's points place equal value on goals and assists. As a replacement for points, goals created assigns a lesser value to assists and is calculated in such a way that adding up each player's goals created equals the actual goals that the team scored. The simplest formula is to add a player's goals to his assists, divide the result by the team's average assists per goal, and then divide that result by two. However, Alan Ryder's revised 2006 formula emphasizes first assists over second assists and treats various manpower situations differently.[180] The Hockey Reference version introduced two years later makes goals

---

180. Alan Ryder, "Goals Created," *Hockey Analytics* (blog), October 31, 2006, http://hockeyanalytics.com/2006/10/goals-created/.

worth twice as much as assists and then adjusts the result so that it adds up to the team's total goals.[181]

GOALS OVER AVERAGE (GOA)

See goals saved above average (below).

GOALS SAVED ABOVE AVERAGE (GSAA)

GSAA is the number of goals prevented relative to the league average, and it is calculated by multiplying the league-average save percentage by the number of shots the goalie faced and then subtracting the goals actually allowed. The modern version was introduced and named by Kurt R. in April 2013, but it's an old concept that was previously introduced as GOA by Phil Myrland in April 2009 and SOAG by Cam Charron in April 2012.[182]

GOAL SUPPORT

Another idea that I shamelessly lifted from baseball in April 2014, goal support is simply the number of goals a team scores in games that a particular goalie started.[183]

GOALS VERSUS SALARY (GVS)

Introduced by me in August 2009, GVS is a player's GVT measured not against the threshold of a replacement-level player, but against the contribution a team would receive from a player with the same cap hit. It's calculated as the player's GVT minus expected GVT, which is the player's cap hit above league minimum, times an annual multiplier.[184] More recent variations might use WAR instead of GVT.

GOALS VERSUS THRESHOLD (GVT)

One of hockey's earliest catch-all statistics, it was introduced by Tom Awad in 2003 as goals versus average, before being re-engineered to

181. "Glossary," Hockey Reference, July 2008, https://www.hockey-reference.com/about/glossary.html.

182. Kurt R., "Using Adjusted Save Percentage, Let's Find the NHL's Best Goalie," *Broad Street Hockey* (blog), April 29, 2013, http://www.broadstreethockey.com/2013/4/29/4280910/nhl-goalie-stats-2013-adjusted-save-percentage; Phil Myrland, "Estimating 1970s Save Percentages," *Brodeur Is a Fraud* (blog), April 5, 2009, http://brodeurisafraud.blogspot.ca/2009/04/estimating-1970s-save-percentages.html; Cam Charron, "Vancouver Goaltenders: A Historical Comparison," *Canucks Army* (blog), April 27, 2012, http://canucksarmy.com/2012/4/27/luongo-was-vancouvers-first-good-goaltender.

183. Rob Vollman, "NHL Goalies 2013–14," Hockey Abstract, April 15, 2014, http://www.hockeyabstract.com/testimonials/nhlgoalies2013–14.

184. Rob Vollman, "Howe and Why: The NHL's Top Values," *Hockey Prospectus* (blog), August 10, 2009, http://www.hockeyprospectus.com/puck/article.php?articleid=240.

measure a player's contributions in terms of goals scored or prevented relative to the threshold of a replacement-level player.[185] Since it uses traditional counting statistics, it can be calculated for virtually any league or era.

GRIT

Introduced by Sportsnet in 2013, grit is the sum of a team's penalty minutes, hits, fights, and blocked shots.[186]

HERO CHART

Originally and briefly known as WARRIOR charts, HERO charts were introduced by Domenic Galamini in February 2015 as a visual representation of a three-year weighted average of a player's key metrics and where that places him on a league-average depth chart.[187]

HEXTALLY CHART

The shot maps introduced for *War on Ice* by Andrew Thomas and Sam Ventura in September 2014 were based on BallR, which are Kirk Goldsberry's Interactive NBA Shot Charts. They were named Hextally charts because they are hex-based, rather than point-based or heat-based, and as an homage to legendary ex-Flyers goalie Ron Hextall.

HISTORICAL PLAYER PROJECTION

In 1986, Bill James introduced similarity scores as a way to statistically calculate the difference between players, which paved the way for a method of projecting a player's future performance as a weighted and era-adjusted average of similar players from the past.[188] These ideas eventually found their way into other sports, including my own hockey

185. Tom Awad, "Numbers on Ice: Understanding GVT, Part 1," *Hockey Prospectus* (blog), July 30, 2009, http://www.hockeyprospectus.com/puck/article.php?articleid=233.

186. Derek Zona, "Sportsnet's Grit Chart Measures the Importance of Grit & Winning," *Copper & Blue* (blog), October 31, 2013, http://www.coppernblue.com/2013/10/31/5052674/sportsnets-grit-chart-measures-the-importance-of-grit-winning.

187. Garret Hohl, "Galamini's WARRIOR Charts: How to Use Them with a Winnipeg Jets Example," *Jets Nation* (blog), February 5, 2016, http://jetsnation.ca/2016/2/5/galamini-s-warrior-charts-how-to-use-them-with-a-winnipeg-jets-example/; Domenic Galamini, "HERO Charts – Forwards," *Own the Puck* (blog), February 16, 2015, http://ownthepuck.blogspot.ca/2015/02/horizontal-evaluative-rankings-optic.html.

188. Bill James, *The Bill James Baseball Abstract* (Ballantine Books, 1986).

model introduced in March 2009, which was flushed in more detail in *Hockey Abstract* and led to prospect projection tools like PCS and DEV.[189]

HIT

Individual hits thrown and taken, whether or not they separated the carrier from the puck, have been recorded by the NHL since the 1997–98 season. Since it is a judgment that appears to lack a precise definition, the results can vary from one set of scorekeepers to another and from one season to another.

HOCKEY ANALYSIS RATING (HARO, HARD, HART)

Introduced by David Johnson in December 2010, HARO, HARD, and HART are expressed as a percentage and represent the extent to which a player boosts (or hurts) a team's offensive, defensive, or total shot-based metrics relative to the expectation given his average teammates and opposition.[190] Given the effort involved in calculating them, they were discontinued in 2016.

HOME-PLATE SAVE PERCENTAGE (HP SV%)

Some of the most dangerous shots are taken inside hockey's home-plate area, which goes from the goalposts to the faceoff dots to the top of the faceoff circles and across. In August 2014, I introduced a stat that measures a goalie's save percentage in shots taken exclusively inside this home-plate area.[191]

ICE TRACK

Sporting Charts introduced their Ice Track (sportingcharts.com/nhl /icetrack) heat-based shot map in November 2013 as a way to visualize the location of a player's individual on-ice shots and individual goals.

ICING

The number of team icings that occur when a player is on the ice, both for and against, were included in the Extra Skater website hosted by Darryl Metcalf in the 2013–14 season.

---

189. Rob Vollman, "Howe and Why: Career Projections for Rookies," *Hockey Prospectus* (blog), March 18, 2009, http://www.hockeyprospectus.com/puck/article.php?articleid=47; Rob Vollman, "Historical Player Projections," *Hockey Abstract* (2013), 183–188.

190. David Johnson, "Introducing New Stats Site," *Hockey Analysis* (blog), December 15, 2010, http://hockeyanalysis.com/2010/12/15/introducing-new-stats-site/.

191. Rob Vollman, "Goaltending Analytics Revisited," *Rob Vollman's Hockey Abstract 2014* (CreateSpace Independent Publishing Platform, 2014), 83–111.

IDEAL POINT (OR IDEAL GOAL OR IDEAL ASSIST)

The first era-adjusted stats weren't made to modern standards, but to the league's average throughout history. In the 1986 *Hockey Compendium*, Jeff Z. Klein and Karl-Eric Reif defined ideal points (or goals or assists) as the number of points a player would have scored if the player's ice time, the number of assists per goal, and the number of goals per game were all equal to the league's average throughout its history.[192]

INDIVIDUAL POINTS PERCENTAGE (IPP)

IPP is the percentage of all a team's on-ice points on which a player recorded a goal or an assist, and it is usually calculated using only even-strength situations. It was originally introduced by me as even-strength points percentage in my July 2008 annual super-spreadsheet, which was based on the presence statistic, which was based on all of a team's scoring, not just what occurred while the player was on the ice.[193] It was later renamed and popularized by Scott Reynolds in October 2012, based on some work by Tyler Dellow from November 2009.[194]

INDIVIDUAL SCORING DOMINANCE (ISD)

ISD was introduced by Jeff Z. Klein and Karl-Eric Reif in 1986 as a measure of a player's scoring relative to the league average. It is calculated as the player's goals (or assists) per game divided by the average team's goals (or estimated assists) per game that season.[195]

INDUCTINATOR

In 1995, Bill James introduced the Hall of Fame monitor to predict which baseball players would make the Hall of Fame by using similarity scores to compare them to those who had been selected in the past.[196] Inspired by that idea, Iain Fyffe introduced the Inductinator to

192. Jeff Z. Klein and Karl-Eric Reif, "Goal-Scorers and Playmakers," *The Klein and Reif Hockey Compendium*, 87–95.

193. Rob Vollman, "Even-Strength Points Percentages," HAG_list, Yahoo Group, July 17, 2008, https://groups.yahoo.com/neo/groups/HAG_list/conversations/topics/1671.

194. Scott Reynolds, "Individual Point Percentages for 2011–12," *NHL Numbers* (blog), October 12, 2012, http://nhlnumbers.com/2012/10/12/individual-point-percentage-for-2011–12; Tyler Dellow, "Points v. Scoring," *Battle of Alberta* (blog), November 27, 2009, http://battleofalberta.blogspot.ca/2009/11/points-v-scoring.html.

195. Jeff Z. Klein and Karl-Eric Reif, "Goal-Scorers and Playmakers: Our Forefathers Had No Other Books but the Score and the Tally," *The Klein and Reif Hockey Compendium*, 80–82.

196. Bill James, chapter 7 in *The Politics of Glory: How the Baseball's Hall of Fame Really Works*, (Macmillan, 1995).

hockey in June 2010 to determine the implicit standards to be voted into the Hall of Fame. The Inductinator divides players by position and era and assigns points for various achievements, and any player exceeding 100 points is expected to be chosen for the Hall of Fame.[197]

K

Introduced by Emmanuel Perry in October 2016, K is a catch-all statistic that uses shot rates, shot quality, penalty rates, and contextual states to measure a player's value in goals relative to a player of no impact.[198]

KRACH RATING

With origins in the 1990s, Ken Butler's rating for American College Hockey (KRACH) is an objective way of ranking teams in leagues with unbalanced schedules using the Bradley-Terry system. It is meant to replace RPI. They are also known as Z-ratings.[199]

LEVERAGE

Micah Blake McCurdy introduced the leverage statistic (briefly known as pressure) in January 2016, which was intended to measure the offensive and defensive importance of a given moment in a game, as calculated by the swing in how many points each team can expect to earn or lose if they score or allow the next goal.[200] In practice, it can be used to identify which players are most often used in key situations.

LOCATION-ADJUSTED EXPECTED GOALS PERCENTAGE (LAEGAP)

See expected goals (page 267).

LOCATION-ADJUSTED STATISTICS

Shot-based metrics like save percentage and Corsi don't account for the quality of the shots taken. Location-adjusted stats use the coordinates in NHL game files to weigh the values of each shot differently. Quality-adjusted stats also provide additional information about the shot. The term *location-adjusted* also sometimes refers to rink-adjusted or venue-adjusted.

---

197. Iain Fyffe, "Up and Coming, Developing the Inductinator," *Hockey Prospectus* (blog), June 14, 2010, http://www.hockeyprospectus.com/puck/article.php?articleid=585.

198. Emmanuel Perry, "Composite Tailored Regression Modeling for Evaluative Ratings in Professional Hockey," (PDF) Corsica Hockey, June 16, 2016, http://corsica.hockey/misc/K_Manuscript.pdf.

199. John Whelan, "KRACH Frequently Asked Questions," College Hockey News, http://www.collegehockeynews.com/info/?d=krach.

200. Micah Blake McCurdy, "Leverage," HockeyViz, August 5, 2016, http://hockeyviz.com/txt/leverage.

LUCK-NEUTRAL STANDINGS

Introduced by me in August 2013, luck-neutral standing is meant to identify, capture, and remove the impact of random variation on league standings.[201] It is loosely based on the team luck index.

MANPOWER-ADJUSTED STATISTICS

See situation-adjusted statistics (page 297).

MARGINAL GOAL

Based on baseball's marginal runs concept included in the win shares system developed by Bill James, marginal goals is an era-adjusted stat used in hockey's catch-all statistics. It is calculated as a team's goals minus one-half the league average. (Note that the point shares system uses 7/12 instead of 1/2 the league average.)

MERCAD (5V5 ADJGSAA/60)

Going all the way back to SQSNV in 2004, numerous goaltending metrics have compared a goalie's actual results to how a league-average goalie would have fared against the same volume and quality of shots. In August 2015, Nick Mercadante put his own spin on these metrics by measuring it in terms of goals saved per 60 minutes, as opposed to a save percentage.[202] In July 2016, Ian Fleming introduced a version of Mercad that uses a different model to account for shot quality, which uses all shot attempts instead of just those that reached the net.[203]

MICRO-STATISTICS

A single hockey game can be broken down into thousands of different plays, like passes, zone entries, blocked shots, and turnovers. When counted either by hand, with camera technology, or with chips in player equipment and/or the puck, these events are known as micro-stats.

MISSED SHOT

One of the RTSS statistics rolled out by the NHL, a missed shot is when a player's undeflected shot misses the net, or would have missed the net if the goalie hadn't made the save, in the opinion of the

---

201. Rob Vollman, "Who Will Finish First Next Year?," *Hockey Abstract* (2013), 76–84.

202. Nick Mercadante, "Goalies Are Voodoo . . . But Improving Comparative Analysis Tools Can Help," *Blue Shirt Banter* (blog), August 12, 2015, http://www.blueshirtbanter.com/analytics/2015/8/12/9136611 /goalies-are-voodoo-but-improving-comparative-analysis-tools-can-help.

203. Ian Fleming, "AdjGSAxGA/60 – Another Different Look at Goaltending," *NHL Numbers* (blog), July 30, 2016, http://nhlnumbers.com/2016/7/30/adjgsaxga-60-a-nother-different-look-at-goaltending.

scorekeeper. Recently, the league has even kept track of which ones hit the post, hit the crossbar, go wide to one side, or go over the net.

NEILSON NUMBER

A version of plus/minus that was popularized by David Staples in 2007 and whose origins trace back to coach Roger Neilson in the 1970s. A player's Neilson number is the number of scoring chances in which the player was directly involved minus those allowed.[204]

NET GOALS POST FACEOFF (NGPF OR NSPF FOR SHOTS)

Instead of measuring faceoff success with a scorekeeper's opinion of who touched the puck first, Craig Tabita introduced a more meaningful and objective method in January 2015.[205] NGPF and NSPF are the number of goals and unblocked shot attempts that occur within 10 seconds of a faceoff, which can be more reflective of which side truly won. Typically, the data is broken down by manpower situation and the zone in which the faceoff occurred.

NET PENALTY DIFFERENTIAL (NPD)

Drawing penalties is a highly undervalued skill. Introduced by me in September 2013, a player's NPD is the number of penalties drawn minus those taken relative to a league-average player in the same position who was assigned the same ice time in each manpower situation.[206]

NHL EQUIVALENCY (NHLE)

Can a player's scoring in other leagues provide insight into how he might perform in the NHL? Based on the Major League Equivalency (MLE) statistic developed by Bill James in 1985, Gabriel Desjardins introduced the NHLe statistic to hockey in December 2004.[207] It is an estimate of how many points a player would score in the NHL based on the player's scoring in another league and the scoring past players

---

204. David Staples, "Frequently Asked Questions on Neilson Numbers," *Edmonton Journal*, September 4, 2012, http://edmontonjournal.com/sports/hockey/nhl/cult-of-hockey/frequently-asked-questions-on-neilson-numbers.

205. Craig Tabita, "Redefining Face-off Success Using Shot Data," *Hockey Prospectus* (blog), January 31, 2015, http://www.hockeyprospectus.com/redefining-face-off-success-using-shot-data/.

206. Rob Vollman, "Los Angeles Kings," *Hockey Prospectus 2013–14*, 167–169.

207. Bill James, *The Bill James Baseball Abstract*, 1985; Gabriel Desjardins, "League Equivalencies," *Hockey Analytics* (PDF on blog), December 2004, http://hockeyanalytics.com/Research_files/League_Equivalencies.pdf.

who also came from that league retained when they moved up to the NHL.

OCTOPLOT

An octoplot is a graph of a team's performance, both offensive and defensive, based on the four possible outcomes of a shot attempt: a goal, a save, a blocked shot, or a miss. It was introduced by Mike Morris in September 2015.[208]

OFFENSIVE PROFILE CHART

One of the total performance charts I introduced for *Hockey Prospectus 2013–14*, a team's offensive profile chart maps each player's shots per 60 minutes on the horizontal axis and set-up passes per 60 minutes on the vertical axis. Sized circles represent even-strength scoring rate.[209]

OFFENSIVE ZONE START PERCENTAGE (OZ%)

See zone start percentage (page 307).

ONE-STAT ARGUMENT

Most commonly applied to goaltenders and even-strength save percentage, the one-stat argument posits that the performance of a team or player can be best evaluated using a single statistic.

ON-ICE AND OFF-ICE STATISTICS

Measuring how a team performs with a particular individual on and off the ice forms the base components used in relative statistics for that player. Its origins go back at least as far as Gabriel Desjardins's *Behind the Net* website, which was launched in 2007.

OSCAR

See Cordelia (page 262).

OVERTIME GOALS (OTG)

This is quite simply the number of goals scored in overtime.

PACE FACTOR

In basketball, the pace of a game is calculated as the number of possessions per game. In hockey, it is calculated as the number of shot attempts per 60 minutes.

PASSING PROJECT

The idea of tracking a game's passes goes back to Lloyd Percival in the

---

208. Mike Morris, "Plotting a New Course," *Winging It in Motown* (blog), September 20, 2015, http://www.wingingitinmotown.com/2015/9/20/9359683/plotting-a-new-course (site discontinued).

209. Rob Vollman, "Introducing Our Total Performance Charts," *Hockey Prospectus 2013–14*, x–xv.

1950s and was a standard exercise for coach Anatoli Tarasov in 1963.[210] Unfortunately, NHL play-by-play files do not include this information, leaving us with only estimates like set-up passes. That inspired Ryan Stimson, who first began tracking passes independently in the 2013–14 season, to organize a volunteer project to manually track all passes.[211]

PDO (SPSV%)

Sometimes referred to as the luck statistic because of its tendency to regress back to 100, a team's PDO is calculated by adding its shooting and save percentages together, generally in even-strength situations only, either overall or with only a specific player on the ice. It was named by Timothy Barnes after Brian King, who first suggested it while using the online handle PDO in 2009, and it is based on the work of Barnes and others.

PENALTIES DRAWN/TAKEN

A player's total number of penalty minutes (PIM) is one of the NHL's oldest statistics, but breaking the data down by type and tracking the number of penalties a player has drawn and taken are relatively new developments.

PENALTY SHOT (PS)

Penalty shots were introduced in the PCHA in 1921–22 and to the NHL in 1934–35.

PERCENTAGE OF TEAM SHOTS (%TSH)

This individual statistic introduced by Ben Wendorf in October 2013 measures how much of a team's shots were contributed by a particular player. Similar to the presence statistic, it is calculated by dividing a player's shots by all shots taken by that team in games in which the individual played.[212]

PERCENTAGE STATISTICS

Rather than expressing a particular piece of information in absolute

---

210. David Staples, "The Eternal Search for the Perfect Hockey Stat, Part 8,348," *Edmonton Journal*, February 13, 2009, http://edmontonjournal.com/sports/hockey/nhl/cult-of-hockey/the-eternal-search-for-the-perfect-hockey-stat-part-8348.

211. Ryan Stimson, "2013–14 Devils Passing Review: A Stats Primer," *All About the Jersey* (blog), July 21, 2014, http://www.allaboutthejersey.com/2014/7/21/5899095/a-passing-stats-primer.

212. Ben Wendorf, "New Metric: % of Team Shots (or %TSH); Using This Historical Shooting Contribution Metric to View Craig Simpson's 1987–88," *Hockey Visualized* (blog), October 2, 2013, https://hockeyvisualized.com/2013/10/02/percentage-team-shots-craig-simpson-87-88/.

terms, such as goals, or as a rate, such as goals per 60 minutes, it can sometimes be useful to express data as a percentage. For example, a team's goal percentage would be its goals divided by all on-ice goals, which is the sum of both its goals and those of its opponents.

PERSEVERANCE

One of the first non-traditional goalie stats, perseverance was introduced by Jeff Z. Klein and Karl-Eric Reif in 1986.[213] To calculate it, multiply a goalie's save percentage by 600, add the number of shots faced, and divide the whole thing by 0.6.

PIP

See Cordelia (page 262).

PLAYER CONTRIBUTION (PC)

An early and sophisticated catch-all statistic introduced by Alan Ryder in August 2003, PC is "a method for allocating credit for a team's performance to the individual contributors on a hockey team."[214]

PLAYER RADAR CHART

Introduced by Ryan Stimson in March 2016 based on charts created for soccer by Ted Knutson in January 2014, player radar charts are a visualization of 12 key individual player metrics.[215]

PLAYER USAGE CHART (PUC)

Originally introduced in June 2011 as OZQoC charts—and sometimes called Vollman sledgehammers, player deployment charts, and bubble charts—player usage charts are a graphic portrayal of how a player is being used.[216] These charts show zone start percentage on the horizontal axis and quality of competition or teammates on the vertical axis. Sized, shaded circles show shot-based metrics and/or ice time data. In October 2014, Megan Richardson introduced a variation

213. Jeff Z. Klein and Karl-Eric Reif, "The Goalers," *The Klein and Reif Hockey Compendium*, 138–159.

214. Alan Ryder, "Player Contribution" (PDF), Hockey Analytics, August 2003, http://www.hockeyanalytics.com/Research_files/Player_Contribution_System.pdf.

215. Ryan Stimson, "Introducing Player Radar Charts," *Hockey Graphs*, June 13, 2016, https://hockey-graphs.com/2016/06/13/introducing-player-radar-charts/; Ted Knutson, "Radar Love: The Three Best Players in the World," *Stats Bomb* (blog), January 20, 2014, http://statsbomb.com/2014/01/radar-love-the-three-best-players-in-the-world/.

216. Rob Vollman, "Winnipeg OZQoC Graphs," *Arctic Ice Hockey* (blog), June 20, 2011, http://www.arcticicehockey.com/2011/6/20/2233834/winnipeg-ozqoc-graphs.

of player usage charts that uses each player's relative zone start percentage on the horizontal axis, which makes it easier to compare players between teams, especially those who changed teams.[217]

PLAYMAKING METRIC (PLAY)

As an improvement on using assists to measure an individual's playmaking ability, Brian Macdonald introduced the PLAY metric in July 2013, which is based on both shots and goals and accounts for the strength of teammates.[218]

PLAYOFF OUTPUT PROJECTION (POP)

Dominik Luszczyszyn developed POP in August 2014 to predict how teams would perform in the playoffs. He wanted to improve on straight-up shot-based metrics.[219] The current formula is to add a team's score-adjusted weighted shots to half of its special teams index in a 0.78 to 0.22 ratio.

PLUS/MINUS (+/-)

Believed to have been formally co-developed by Allan Roth and Dick Irvin of the Montreal Canadiens in the early 1950s, or possibly by Scotty Bowman in the early 1960s, plus/minus was officially recorded no later than the 1967–68 season.[220] It is the number of goals a team scores while a player is on the ice minus those allowed, not counting power-play goals (but it does include empty-net goals and penalty shots).

POINT ALLOCATION (PA)

Hockey's first catch-all statistic was introduced by Iain Fyffe in April 2002.[221] The most notable aspect of the calculation is how a player's

217. Megan Richardson, "Relative Zone Start Charts: Better Visualizing Coaches' Deployment," Progressive Hockey, October 5, 2014, http://www.progressivehockey.com/2014/10/relative-zone-start-charts-better.html.

218. Brian Macdonald "PLAY," *Greater Than Plus/Minus* (blog), July 2013, http://www.greaterthanplusminus.com/p/play.html.

219. Dominik Luszczyszyn, "Why Possession Isn't Everything," *Hi This is Dom* (blog), August 27, 2014, https://hithisisdom.wordpress.com/2014/08/27/why-possession-isnt-everything/.

220. David Staples, "Just How Accurate Is Official NHL Plus-Minus? One More Nail in the Coffin," *Edmonton Journal*, April 29, 2014, http://edmontonjournal.com/sports/hockey/nhl/cult-of-hockey/just-how-accurate-is-official-nhl-plus-minus-one-more-nail-in-the-coffin; Mike Commito, "The History and Future of Hockey's Most Polarizing Statistic," Sportsnet, June 24, 2016, http://www.sportsnet.ca/hockey/nhl/just-doesnt-add-history-future-plus-minus/.

221. Iain Fyffe, "Puckering Archive: Point Allocation (09 April 2002)," *Hockey Historysis* (blog), September 19, 2014, http://hockeyhistorysis.blogspot.ca/2014/09/puckerings-archive-point-allocation-09.html.

defensive allocations are based on whatever portion of his ice time can't be explained by his scoring.

POINT SHARES (PS)

Point shares is a catch-all statistic introduced by Justin Kubatko in May 2011.[222] It is based on Bill James's win shares baseball statistic and hockey's PA, PS, and GVT statistics. This system assigns a team's offensive and defensive contributions among its players, in terms of points in the standings. Neil Paine once referred to a version that was calculated above replacement level (PSAR), but there's no other record of it.[223]

POSSESSION STATISTICS

Technically, possession stats are a measurement of time that a team spends in control of the puck, sometimes recorded exclusively in the offensive zone. Ever since shot-based metrics were discovered to be a close proxy by Timothy Barnes in October 2007, the two terms are sometimes used interchangeably.[224]

PRESENCE

Presence was introduced by Karl-Erik Reif and Jeff Z. Klein in 1986 as a measure of "how much a player contributes to his own team's attack."[225] It is calculated as a player's own scoring pro-rated to a complete season divided by how many goals his team scored beyond his own.

PRESSURE

See leverage (page 276).

PRIMARY SHOT CONTRIBUTIONS (PSC)

First defined by Ryan Stimson in January 2016, a primary shot contribution is the sum of shot attempts plus passes that directly lead to shot attempts.[226]

---

222. Justin Kubatko, "Calculating Point Shares," Hockey Reference, May 2011, http://www.hockey-reference.com/about/point_shares.html.

223. Neil Paine, "The Amazing Longevity of Jaromir Jagr," *Five Thirty Eight* (blog), February 26, 2015, https://fivethirtyeight.com/features/the-amazing-longevity-of-jaromir-jagr/.

224. Timothy Barnes, "Corsi Numbers," *Irreverent Oilers Fans* (blog), October 2007, http://vhockey.blogspot.ca/2007/10/corsi-numbers.html (no longer available).

225. Jeff Z. Klein and Karl-Eric Reif, "Goal-Scorers and Playmakers," *The Hockey Compendium*, 83–84.

226. Ryan Stimson, "Redefining Shot Quality: One Pass at a Time," *Hockey Graphs* (blog), January 27, 2016, https://hockey-graphs.com/2016/01/27/redefining-shot-quality-one-pass-at-a-time/.

PROJECTINATOR

Devised by Iain Fyffe in October 2009, the Projectinator was the first system to predict a prospect's NHL future based on his totals in junior hockey.[227] The latest version, which is explained in more detail in *Stat Shot*, starts with a player's scoring totals, takes age and league quality into consideration, adjusts it for size and number of games played, and projects how well the player will perform in the NHL in terms of points and/or GVT.[228]

PROSPECT COHORT SUCCESS (PCS)

PCS is a historical player projection system designed by Cam Lawrence and Josh Weissbock in May 2015. It is used to calculate a prospect's chances of making the NHL and to predict how effective he will be if he does make it.[229] To accomplish this, PCS combs though historical data to form a cohort of players of a similar size with similar era-adjusted scoring at the same age as the target player, and then it adjusts for league quality. A prospect's chances of NHL success are based on the percentage of those in the cohort who played at least 200 games, among which an average NHL scoring rate is calculated.

PULL

First recorded by me in April 2014, pulls are the number of games in which the starting goalie was pulled from the game, either due to injury or poor play, but not for an extra attacker or a delayed penalty.[230]

PYTHAGOREAN WINNING PERCENTAGE

Developed for baseball by Bill James, a team's expected winning percentage is calculated as the square of their runs scored divided by the sum of the square of their runs scored and runs allowed. It was first adapted to hockey by Chris Apple and Marc Foster in January 2002, using goals instead of runs, and further developed by James. J. Cochran

227. Iain Fyffe, "Up and Coming, Projecting Defencemen," *Hockey Prospectus* (blog), October 5, 2009, http://www.hockeyprospectus.com/puck/article.php?articleid=294.

228. Iain Fyffe, "What Do a Player's Junior Numbers Tell Us?," *Stat Shot*, 75–120.

229. Josh Weissbock, "Draft Analytics: Unveiling the Prospect Cohort Success Model," *Canucks Army* (blog), May 26, 2015, http://canucksarmy.com/2015/5/26/draft-analytics-unveiling-the-prospect-cohort-success-model.

230. Rob Vollman, "NHL Goalies 2013–14," Hockey Abstract, April 15, 2014, http://www.hockeyabstract.com/testimonials/nhlgoalies2013–14.

and Rob Blackstock in 2009, and by Kevin Dayaratna and Steven J. Miller in 2013.[231] Exponents other than two are sometimes used.

QUALITY-ADJUSTED STATISTICS

Going back as far as Alan Ryder's SQSNV in 2004, save percentages and shot-based metrics are sometimes adjusted for the combined impact of not just the location of a shot but its timing, its type, and other quality-related circumstances.

QUALITY OF COMPETITION (QOC)

Opponents against whom a player is matched up can have a big impact on most stats, which is why Gabriel Desjardins introduced quality of competition metrics in December 2006.[232] QoC is calculated as a time-weighted average of an individual player's opponents for a chosen statistic. Initially, Desjardins chose adjusted plus/minus per 60 minutes for this weighted average, but he settled on shot-based metrics some time in 2010, which has essentially become the standard. However, quality of competition metrics could be based on any stat, such as scoring rate, as Jonathan Willis demonstrated in 2008; average ice time, which Eric Tulsky first proposed in August 2012; catch-all statistics; or expected goals.[233] There have also been entirely new quality of competition metrics introduced, like WoodMoney and Star Percentage, but they mostly follow the same basic premise.

QUALITY OF TEAMMATES (QOT)

See quality of competition (page 285–286), but substitute "linemates" for "opponents."

QUALITY OF VICTORY

Introduced by Jeff Z. Klein and Karl-Eric Reif in their 1986

231. Chris Apple and Marc Foster, "Glancing into the Crystal Ball: Playoffs Projections Based on Pythagorean Performance," *Sports Illustrated*, 12 January 2002, http://sportsillustrated.cnn.com/statitudes /news/2002/01/09/just_stats/; James J. Cochran and Rob Blackstock, "Pythagoras and the National Hockey League," *Journal of Quantitative Analysis in Sports* 5.2 (2009); Kevin D. Dayaratna and Steven J. Miller, *The Pythagorean Won-Loss Formula and Hockey*, Williams University, 2013, http://web.williams .edu/Mathematics/sjmiller/public_html/math/papers/DayaratnaMiller_HockeyFinal.pdf.

232. Gabriel Desjardins, "Behindthenet Ratings: Quality of Competition," *Behind the Net* (blog), March 2007, http://www.behindthenet.ca/qual_comp.html (no longer available).

233. Jonathan Willis, "The Most Interesting Thing I've Come Across in Some Time," *Copper & Blue* (blog), December 30, 2008, http://www.coppernblue.com/2008/12/most-interesting-thing-ive-written-in. html; Eric Tulsky, "A Competition Metric Based on Ice Time," *NHL Numbers* (blog), August 16, 2012, http://nhlnumbers.com/2012/8/16/a-competition-metric-based-on-ice-time.

Compendium, quality of victory is the percentage difference between the scoring leader's goals per game (or assists or points) and the scorer who is in second place.[234]

QUALITY START (QS)

Inspired by the pitching statistic developed by John Lowe for baseball in December 1985, I introduced quality starts to hockey in March 2009 as a measurement of how often a goalie posted a single-game save percentage above the league average.[235] Some versions of quality starts also have an adjustment for low shot volumes.

RATE STATISTICS (/GP /60)

Rate statistics are commonly used to compare teams or players who have had a different number of opportunities to produce. They are created by taking a counting statistic, like goals, and dividing it by the opportunities to create them, like shots, games, or minutes played.

RATING PERCENTAGE INDEX (RPI)

Developed for all NCAA sports in 1981, RPI is a way to rank teams based on the weighted sum of their win-loss record and their strength of schedule (SOS) in a one-to-three ratio.

REALLY BAD START (OR BLOWN START)

A term introduced in tandem with quality starts, a really bad start occurs in any game in which the starting goalie fails to stop at least 85% of shots faced.

REAL-TIME SCORING STATISTICS (RTSS)

Beginning as early as the 1997–98 season in some cases, the NHL began tracking certain micro-stats, such as hits, blocked shots, give-aways, and takeaways. Since each stat includes a subjective element and appears to lack a precise definition, they may be recorded differently from one set of scorekeepers and/or season to the next.

REBOUND

Since rebounds are not officially recorded in game files, they were first statistically defined by Alan Ryder in January 2004 as any shot that

---

234. Jeff Z. Klein and Karl-Eric Reif, "Goal-Scorers and Playmakers," *The Klein and Reif Hockey Compendium*, 76–81.

235. John Lowe, *Philadelphia Inquirer*, December 26, 1985; Rob Vollman, "Howe and Why: Quality Starts," *Hockey Prospectus* (blog), March 25, 2009, http://www.hockeyprospectus.com/puck/article.php?articleid=54.

occurs within 25 feet (7.6 m) and two seconds of another shot, without an intervening faceoff or other event.[236]

REGRESSED EVEN-STRENGTH BLOCKED SHOTS (RESB%)

See even-strength blocked shots percentage (page 267).

RELATIVE PLUS/MINUS (RPM, +- REL)

The most common adjusted form of plus/minus is to calculate it relative to the strength of a player's team, much as Klein and Reif first did in 1986.[237] The variations in use today were introduced by Gabriel Desjardins in 2006, Tom Awad in April 2009, and Scott Reynolds in May 2011.[238]

RELATIVE STATISTICS (REL PREFIX)

Popularized by Gabriel Desjardins on Behind the Net in December 2006, relative stats are a way of measuring a player's performance relative to the rest of his team. They are commonly expressed as the difference between a team's on-ice statistics with that particular player minus their off-ice statistics when that player is on the bench.

RELATIVE ZONE START CHARTS

See player usage chart (page 282).

REPLACEMENT LEVEL

A concept introduced to baseball by Bill James in 1983 and popularized by Keith Woolner in 2001, replacement level is how well a freely available injury call-up would perform.[239] Its first appearance in hockey was in 2003, as part of early catch-all statistics like point allocations, player contribution, and GVT.

RINK-ADJUSTED STATISTICS

Due to the subjective nature of most NHL statistics, some stats can be affected by a systemic scorekeeper bias, as first documented by

236. Alan Ryder, "Shot Quality" (PDF), Hockey Analytics, January 2003, http://hockeyanalytics.com/Research_files/Shot_Quality.pdf.

237. Jeff Z. Klein and Karl-Eric Reif, "Plus/Minus: The Red and the Black," *The Klein and Reif Hockey Compendium*, 96–130.

238. Tom Awad, "Numbers on Ice, Fixing Plus/Minus," *Hockey Prospectus* (blog), April 3, 2009, http://www.hockeyprospectus.com/puck/article.php?articleid=64; Scott Reynolds, "CHL Relative +/- and NHL Success," *Copper & Blue* (blog), May 2, 2011, http://www.coppernblue.com/2011/5/2/2148876/chl-relative-and-nhl-success.

239. Brandon Heipp, "A Brief, Incomplete History of Replacement Level," Baseball Prospectus, October 30, 2012, http://www.baseballprospectus.com/article.php?articleid=18790.

Alan Ryder in 2007, with shot locations.[240] By comparing a team's performance at home and on the road, the extent of this bias can be measured and used to make rink adjustments.

ROYAL ROAD

In a December 2014 radio interview, Steve Valiquette introduced the concept of a royal road, which is an imaginary line that divides the ice in half, going directly from the middle of one net to the middle of the other. Shots that follow puck movement across the royal road are believed to be more likely to result in goals.[241]

RUSH SHOTS

Technically, rush shots are those that occur when an attacking team first gains the zone. Since that information isn't recorded in game files, David Johnson introduced the statistical definition of a rush shot in July 2014 as any shot that occurs within 10 seconds of a shot attempt, faceoff, hit, giveaway, or takeaway in the neutral or opposing zone and without an intervening event in the defending zone.[242]

SALARY GENERATOR

In September 2014, Matt Cane introduced a player comparison tool to predict a player's salary based on that of the most similar players.[243] In October 2016, Emmanuel Perry introduced a version with a different player comparison method.[244]

SAT

See Corsi (page 263).

SAVE

Goalies are awarded a save if, in the scorekeeper's judgment, they

240. Alan Ryder, "Product Recall Notice for Shot Quality," Hockey Analytics, 2007, http://hockeyanalytics.com/2007/06/product-recall-notice-for-shot-quality/.

241. Steve Valiquette, "Valiquette: Tracking How Goals Are Scored" (audio clip), TSN 1290 Radio, December 11, 2014, http://www.tsn.ca/radio/winnipeg-1290/valiquette-tracking-how-goals-are-scored-1.157634.

242. David Johnson, "Introducing Rush Shots," *Hockey Analysis* (blog), July 9, 2014, http://hockeyanalysis.com/2014/07/09/introducing-rush-shots/.

243. Matt Cane, "Player Comparison Tool/Salary Generator," *Puck Plus Plus* (blog), September 10, 2014, https://puckplusplus.com/2014/09/10/player-comparison-toolsalary-generator/.

244. Emmanuel Perry, "Hockey and Euclid: Predicting AAV with K-Nearest Neighbours," *Corsica Hockey* (blog), October 31, 2016, http://www.corsica.hockey/blog/2016/10/31/hockey-and-euclid-predicting-aav-with-k-nearest-neighbours/.

stop a shot that otherwise would have gone in. Saves can therefore be affected by a scorekeeper's bias. In the NHL, saves were first recorded for individual goalies in the 1955–56 season, but the earliest records for these stats go back at least as far as the Manitoba senior league of the mid-1910s.

SAVE CHARTS (SAV3)

Inspired by HERO charts, Ian Fleming introduced SAVE charts in May 2016 as a visual way to compare goaltenders to each other and the league average in a variety of different areas. Save charts can be found at Dispelling Voodoo (dispellingvoodoo.com/save-chart).

SAVE PERCENTAGE (SV%)

Officially recorded starting in the 1982–83 season, but unofficially in the 1950s in the NHL and as far back as the 1910s in other leagues, save percentage is the number of saves a goalie makes divided by shots faced. For whatever reason, it was multiplied by 100 and called efficiency by Jeff Z. Klein and Karl-Eric Reif in 1986.[245]

SAVE PERCENTAGE+ (SV%+)

Inspired by baseball's ERA+ statistic, Cam Charron introduced an era-adjusted version of save percentage in April 2012, which allows you to measure a goalie's performance relative to league average and then compare it to other seasons.[246] As described by a blogger named Kurt R. in July 2012, SV%+ is calculated as follows: subtract the league-average save percentage from one, divide by one minus the goalie's save percentage, and multiply by 100.[247] It was added to Hockey Reference in October 2013 but was soon redeployed as GA%– by swapping the numerator and denominator.[248]

245. Jeff Z. Klein and Karl-Eric Reif, "The Goalers," *The Klein and Reif Hockey Compendium*, 138.

246. Cam Charron, "Vancouver Goaltenders: A Historical Comparison," *Canucks Army* (blog), April 27, 2012, http://canucksarmy.com/2012/4/27/luongo-was-vancouvers-first-good-goaltender/.

247. Kurt R., "Evaluating Goalies in Context: Adjusted Save Percentage, or SV%+," *Broad Street Hockey* (blog), July 21, 2012, http://www.broadstreethockey.com/2012/7/21/3021129/evaluating-goalies-in-context-adjusted-save-percentage-or-sv.

248. Neil, "SV%+ Calculation Poll," Hockey Reference, October 10, 2013, http://www.sports-reference.com/blog/2013/10/sv-calculation-poll/; Mike, "SV%+ is now GA%–," Hockey Reference, October 17, 2013, http://www.sports-reference.com/blog/2013/10/sv-is-now-ga/.

SAVES OVER AVERAGE GOALTENDER (SOAG)

See GSAA (page 272).

SCHEDULE-ADJUSTED STATISTICS

Since certain team statistics can be influenced by the strength of a team's opponents, Micah Blake McCurdy advanced the notion of making schedule adjustments in November 2014.[249] The concept goes back at least as far as Tore Purdy in November 2010.[250] Some variations also adjust for days of rest between games.

SCORE-ADJUSTED CORSI (OR FENWICK)

The most popular method of accounting for the skewing effect of teams playing to the score may have first been proposed in April 2010 by Gabriel Desjardins and then analyzed in more detail by Eric Tulsky in January 2012 and further analyzed with even more modern adjustments by Micah Blake McCurdy and Emmanuel Perry in October 2014.[251] It involves calculating the impact that different score states can have and calculating a team's Corsi relative to those expectations.

SCORE-ADJUSTED PDO

In October 2014, a blogger known as Pet Bugs introduced a score-adjusted PDO, which adjusts for changes in average team shooting and save percentages when teams are chasing or protecting a late lead.[252]

SCORE-ADJUSTED STATISTICS

Established at least as far back as Phil Myrland in January 2009, shot-based (and other) metrics can be affected by the score, especially in the third period.[253] Teams protecting a late lead tend to sit back

249. Micah Blake McCurdy, "Schedule Adjustment for Counting Stats," *Hockey Graphs* (blog), November 25, 2014, https://hockey-graphs.com/2014/11/25/schedule-adjustment-for-counting-stats/.

250. Tore Purdy, "Adjusted Corsi Update and Goalposts," *Objective NHL* (blog), November 30, 2010, http://objectivenhl.blogspot.ca/2010/11/adjusted-corsi-update-and-goalposts.html.

251. Gabriel Desjardins, "Corsi and Score Effects," *Arctic Ice Hockey* (blog), April 13, 2010, http://www.arcticicehockey.com/2010/4/13/1416623/corsi-and-score-effects; Eric Tulsky, "Adjusting for Score Effects to Improve Our Predictions," *Broad Street Hockey* (blog), January 23, 2012, http://www.broadstreethockey.com/2012/1/23/2722089/score-adjusted-fenwick; Micah Blake McCurdy, "Better Way to Compute Score-Adjusted Fenwick," HockeyViz, October 2014, http://hockeyviz.com/txt/senstats.

252. Pet Bugs, "Score-Adjusted PDO," *NHL Numbers* (blog), October 24, 2014, http://nhlnumbers.com/2014/10/24/score-adjusted-pdo.

253. Phil Myrland, "Playing to the Score," *Brodeur Is a Fraud* (blog), January 5, 2009, http://brodeurisafraud.blogspot.ca/2009/01/playing-to-score.html.

and take fewer shots, while those chasing the game will open up, take more risks, and put more pucks on the net. Early statistics overcame this problem by using only data from close games, but modern metrics measure and adjust for these score effects.

SCORE CLOSE STATISTICS

See close game statistics (page 262).

SCORE-STATE DEPLOYMENT CHARTS

By December 2015, Micah Blake McCurdy had introduced team-by-team score-state deployment charts, which are a graphical representation of which players get the greatest percentage of a team's even-strength ice time in seven different score situations, including when the score is tied, when they are trailing by one, two, or three or more goals, and when leading by the same.[254]

SCORING CHANCE (SC)

Traditionally, scoring chances are defined as any shot taken from within the home-plate area, including those that missed the net but not those that were blocked. The definition also allowed for some scorekeeper discretion if there was dangerous puck movement before the shot or if the goalie was screened. In December 2014, Andrew Thomas and Sam Ventura statistically formalized that definition to include all unblocked shots from within the home-plate area as well as unblocked rebounds and rush shots from the outside and blocked shots in the slot.[255]

SCORING CHANCE PROJECT

Using the traditional definition of a scoring chance, Dennis King manually tracked scoring chances for the Edmonton Oilers in 2008–09, which inspired several more analysts to do likewise in the following seasons, which led to the community's first volunteer manual tracking project in 2011–12.[256]

---

254. Rob Vollman, "Numbers Defend Karlsson's Case for the Norris Trophy," National Hockey League, December 17, 2015, https://www.nhl.com/news/numbers-defend-karlssons-case-for-norris-trophy/c-793148.

255. Andrew Thomas, "NEW: Defining Scoring Chances," *War on Ice* (blog), December 2014, http://blog.war-on-ice.com/new-defining-scoring-chances/index.html.

256. Derek Zona, "Join the Scoring Chance Project," *Copper & Blue* (blog), September 6, 2011, http://www.coppernblue.com/2011/9/6/2407259/join-the-scoring-chance-project.

SCORING NETWORK CHART (OR GOAL AND ASSIST NETWORK)

With early variations designed by November 2015, Micah Blake McCurdy introduced scoring network charts, which are a team-level graphical representation of the goals and primary assists scored by individual players in each manpower situation, with a link to the teammates who are most frequently involved in them.[257]

SCRAMBLE SHOT

In January 2004, Alan Ryder defined a scramble shot as "a shot of less than six feet [1.8 m] that was neither a rebound nor an empty net goal."[258]

SEAL-ADJUSTED SCORING

Introduced by Garret Hohl in June 2016, SEAL-adjusted scoring is a system that measures a prospect's scoring level.[259] It adjusts for league quality like the classic NHLe, adjusts for era and age like the version introduced by Rhys Jessop in June 2014, and reduces the weight of secondary assists.[260]

SET-UP PASSES (SP)

Defined as any pass that results in a shot attempt, this method to estimate set-up passes was introduced by me in December 2012.[261] As detailed in *Hockey Abstract*, set-up passes are estimated by dividing a player's primary assists by the team's on-ice shooting percentage in each manpower situation.[262]

SHIFT

Officially recorded by the NHL since the 1997–98 season, a shift begins when a player gets on the ice and ends when he leaves it. In the

257. Micah Blake McCurdy (@IneffectiveMath), "Detroit scoring network," Twitter, November 27, 2015, 4:19 p.m., https://twitter.com/ineffectivemath/status/670396446793953280.

258. Alan Ryder, "Shot Quality," Hockey Analytics, January, 2014, http://hockeyanalytics.com/2004/01/isolating-shot-quality/.

259. Garret Hohl, "SEAL-Adjusted Scoring and Why It Matters for Prospects," *Hockey Graphs* (blog), June 15, 2015, https://hockey-graphs.com/2016/06/15/seal-adjusted-scoring-and-why-it-matters-for-prospects/.

260. Rhys Jessop, "Adjusted Draft Year Scoring for CHL Prospects," *That's Offside* (blog), June 15, 2014, http://thats-offside.blogspot.ca/2014/06/adjusted-draft-year-scoring-for-chl.html.

261. Rob Vollman, "Howe and Why: Passes," *Hockey Prospectus* (blog), December 31, 2012, http://www.hockeyprospectus.com/puck/article.php?articleid=1417.

262. Rob Vollman, "Passes," *Hockey Abstract*, (2013), 210–217.

NHL, they first came about when the new rules for offsides in 1929 sped up the game, and they became common practice in the 1930s.

SHIFT CHARTS

A shift chart is a graphical representation of which players were on the ice at which times, including with whom and against whom they were playing. Introduced by the NHL in 2002–03, they were removed for the 2007–08 season, at which point they were only available at various independent hobbyists' sites, before being reintroduced by the NHL in 2010–11.[263]

SHOOTING PLUS SAVE PERCENTAGE (SPSV%)

See PDO (page 280).

SHOOTOUT (SO PREFIX)

Essentially just a penalty shot, the shootout evolved from intermission entertainment to a post-game method of deciding tied hockey games at the 1988 Winter Olympics. Since it was introduced for the 2005–06 season, the NHL has been tracking shootout attempts, goals, and game-deciding goals for players, goalies, and teams.

SHOT (SOG)

A shorthand for shot on goal, a shot is any attacking shot attempt that went into the goal or would have entered the goal if it hadn't been stopped by the goaltender. Blocked shots, missed shots, or saved shots that would have missed the net in the opinion of the scorekeeper are not recorded as shots. Own goals are recorded as shots, for the attacking player who most recently touched or was closest to the puck.

SHOT ASSIST (OR PRIMARY SHOT ASSIST)

This is a manually tracked version of set-up passes and was introduced by Ryan Stimson as part of the Passing Project. It was named as such in November 2015 (at the latest).[264]

SHOT ATTEMPT

Shot attempts include all shots on goal, plus those that were blocked or missed the net. They form the basis of the Corsi statistic.

---

263. An example of an original NHL shift chart from 2002–03 can be found at http://www.nhl.com /scores/htmlreports/20022003/SCH21230.gif. An example of the re-introduced NHL shift chart from 2010–11 can be found at http://www.nhl.com/stats/shiftcharts?id=2010020001.

264. Ryan Stimson (@RK_Stimp), "And here's TOR shot assists for (black) and against (red) w/ Gardiner on the nice. Guesses what side he plays?" Twitter, November 19, 2015, 12:11 p.m., https://twitter.com/ rk_stimp/status/667434971846561793.

SHOT ATTEMPT GENERATION EFFICIENCY (SAGE)

Introduced by Ryan Stimson in May 2014, SAGE is the efficiency with which passes create shots that actually get through to the net.[265] It has a particularly tight correlation with winning. The formula is shots on goal generated from passes divided by all shot attempts generated from passes.

SHOT-BASED METRICS

Early individual shot-based metrics include shots and shooting percentage, for which official counts are currently available back to the 1967–68 NHL expansion, and goalie save percentage, which goes back to 1955–56. Since the popularization of the Corsi statistic in 2006, shot-based metrics typically refer to how a team performs in terms of all shot attempts with a given player on the ice.

SHOT DISTANCE

The NHL began recording more detailed information in its game files with the 2002–03 season, including the exact timing, type, and even coordinates of a shot (unless it was blocked). This information quickly led to location-adjusted statistics.

SHOT MAP

When the NHL first started recording shot locations in 2002–03, it became possible to create heat maps that showed where shots were taken—and the likelihood of scoring from each location—for teams, players, and goalies. These charts now come in different styles and go by different names, such as game shot maps, Ice Track, and Hextally charts.

SHOT QUALITY

One of the field's most controversial metrics, shot quality is a measure of the danger of a shot based on its location and timing. It is used in expected goals stats (see page 267) and quality-adjusted stats (see page 285).

SHOT QUALITY NEUTRAL SAVE PERCENTAGE (SQNSV)

The first quality-adjusted save percentage was devised by Alan Ryder in January 2004, who described SQNSV as "the save percentage one

---

265. Ryan Stimson, "Passing Efficiency of the New Jersey Devils," *Washington Post*, May 8, 2014, https://www.washingtonpost.com/news/fancy-stats/wp/2014/05/08/passing-efficciency-of-the-new-jersey-devils.

would expect with no variation in shot quality from team to team."[266] The calculation involves classifying shots based on type, timing, distance, and manpower situation. After Ryder observed that inaccurate scorekeeping was affecting the results, Chris Boersma introduced a rink-adjusted version in September 2007.[267]

SHOT QUALITY PROJECT

Given the unreliability of shot location data in NHL game files and the absence of information like transitions, passes, and rebounds, Chris Boyle launched the Shot Quality Project in October 2013 to manually compile more complete and reliable data, to be used primarily in goaltending analyses.[268]

SHOT TIDE

In April 2016 (at the latest), Micah Blake McCurdy introduced shot tide charts, which are a rolling average of each team's shots per 60 minutes throughout the course of a game. Man-advantage situations are shaded, and goals are denoted.[269]

SHOT TYPES

The NHL began recording shot types in its game files in 2002–03, including backhands, deflections, slap shots, snap shots, tipped shots, wraparounds, and wrist shots. Apparently without a precise definition, the recorded results tend to vary significantly from one arena's scorekeepers to another's.

SHUTOUTS

A team and/or goalie is credited with a shutout in any game in which the opposing team is prevented from scoring a goal in any fashion except in the shootout. In June 2007, Phil Myrland wrote

---

266. Alan Ryder, "Shot Quality" (PDF), Hockey Analytics, January, 2004, http://hockeyanalytics.com/Research_files/Shot_Quality.pdf.

267. Chris Boersma, "Adjusted Shot Quality Neutral Save Percentage," *Hockey Numbers* (blog), September 14, 2007, http://hockeynumbers.blogspot.ca/2007/09/adjusted-shot-quality-neutral-save.html.

268. Chris Boyle, "Introducing the Shot Quality Project," Sportsnet, October 22, 2013, http://www.sportsnet.ca/hockey/nhl/introducing-the-shot-quality-project/.

269. Micah Blake McCurdy (@IneffectiveMath), "Less that the stars shelled and more that the wild brought everything. Other charts up at hockey.viz.com," Twitter, April 24, 2016, 3:32 p.m., https://twitter.com/IneffectiveMath/status/724365423190380545.

that "shutouts are arbitrary, improperly weighted, strongly team-dependent, and poor measures of goaltender ability."[270]

SIMILARITY SCORES

Bill James introduced similarity scores to baseball in 1986 as a way to statistically calculate the difference between players and to find those who are similar.[271] In hockey, it has been used in historical player projection systems, salary generators, and more.

SIMPLE RATING SYSTEM (SRS)

Introduced to football by Doug Drinen in May 2006, SRS measures how well a team performs in terms of scoring differential relative to the average scoring differential of their opponents.[272] It was imported to hockey in October 2010 (or perhaps earlier).[273]

SITUATION-ADJUSTED SAVE PERCENTAGE

Phil Myrland introduced a version of save percentage adjusted for manpower situation in October 2009.[274] It combines even-strength and short-handed save percentage in a four-to-one ratio and sets power-play save percentage aside.

SITUATION-ADJUSTED STATISTICS

Since the quality and quantity of most events are different in each manpower situation, some stats can be skewed by the varying amounts of time that certain teams or individuals spend playing in particular situations. Some metrics manage that by looking exclusively at one situation, such as even-strength statistics, but situation-adjusted stats take advantage of all available data either by measuring and adjusting for the impact of each manpower situation on the given event or by combining a team's or player's data in each manpower situation into a common ratio.

---

270. Phil Myrland, "Vezina Trophy: The Worthlessness of Shutouts," *Brodeur Is a Fraud* (blog), June 11, 2007, http://brodeurisafraud.blogspot.ca/2007/06/vezina-trophy-worthlessness-of-shutouts.html.

271. Bill James, *The Bill James Baseball Abstract, 1986*.

272. Doug Drinen, "A Very Simple Ranking System," Pro Football Reference, May 8, 2006, http://www.pro-football-reference.com/blog/index4837.html?p=37.

273. Phil Myrland, "SRS," *Brodeur Is a Fraud* (blog), October 5, 2010, http://brodeurisafraud.blogspot.ca/2010/10/srs.html.

274. Phil Myrland, "Situation-Adjusted Save Percentage," *Brodeur Is a Fraud* (blog), October 12, 2009, http://brodeurisafraud.blogspot.com/2009/10/situation-adjusted-save-percentage.html.

SKATER CONTEXT DIAGRAMS

In 2016, Micah Blake McCurdy introduced skater context diagrams, which are a graphical representation of how a player is deployed in relation to the average quality of his linemates and opponents. It is broken down between forwards and defencemen. This quality is measured in terms of average ice time.[275]

SNIFF TEST (OR SANITY CHECK)

A casual phrase I first heard from Iain Fyffe in the early 2000s, a sniff test is when a statistician looks for potential major flaws in a statistic by comparing the results to established opinion.

SPECIAL TEAMS INDEX (STI)

Believed to have been first mentioned by Scotty Bowman while coaching the Montreal Canadiens in the 1970s, special teams index is a team's power-play percentage plus their penalty-killing percentage.

SPECIAL TEAMS PROJECT

In January 2016, Arik Parnass launched the Special Teams Project to manually track the data required to analyze man-advantage play.[276]

SPECIAL TEAMS USAGE CHART

One of the four total performance charts I introduced in *Hockey Prospectus 2013–14*, a special teams usage chart lays out a player's average power-play ice time on one axis and penalty-killing time on the other. Shaded circles represent even-strength ice time.[277]

STAR PERCENTAGE

A new quality of competition metric was hinted at by Tyler Dellow in February 2017, which appears to be based on the percentage of a player's ice time spent against star players.[278]

STRENGTH OF SCHEDULE (SOS)

Used for the RPI rating system, a team's SOS is the weighted sum of

275. The data source for the skater context diagrams is HockeyViz, http://hockeyviz.com/howto/skaterContext.

276. Arik Parnass, "Introducing the Special Teams Project," Special Teams Project, January 13, 2016, http://www.nhlspecialteams.com/blog/2016/1/11/introducing-the-special-teams-project.

277. Rob Vollman, "Introducing Total Performance Charts," *Hockey Prospectus 2013–14*, x–xv.

278. Tyler Dellow (@dellowhockey), "I have a measure of defence QoC I call star percentage. This is what it looks like for the Leafs this year. Draw your own conclusions." Twitter, February 26, 2017, 10:47 a.m., https://twitter.com/dellowhockey/status/835924320232824832.

the average winning percentage of a team's opponents and the average winning percentage of the opponents of those opponents. It is measured in a two-to-one ratio.

STRUCTURE INDEX

Introduced by Matt Cane in August 2017, a team's structure index is a reflection of how consistently their power-play shots occur from set locations.[279]

TAKEAWAYS

Introduced with the other RTSS statistics (see page 287) in 1997–98, takeaways are ostensibly awarded to a defending player whose actions result in his team gaining possession of the puck. However, in the absence of a precise definition, takeaways can be subjective (for example, it seems even the best players steal the puck only once per game), and results can vary from one scorekeeper to another.

TARASOV NUMBERS

Inspired by Anatoli Tarasov's work, David Staples introduced a new plus/minus statistic in February 2009, in which players get points for a "positive contribution in the sequence that leads to a goal," such as a pass, clearance, rush, or shoot in, while getting penalized for an error that resulted in a goal against.[280]

TEAM LUCK INDEX

Introduced by me in August 2013, the team luck index is a measure of how many points a team gained or lost in the standings due to random variation. It includes the impact of injuries, shooting and save percentages, special teams, and the team's record in one-goal games, overtime, and the shootout.[281]

TIME ON ICE (TOI)

First introduced by the NHL in the 1997–98 season, ice time is arguably the most important individual player statistic. It measures a player's performance relative to the rest of his teammates, in the opinion

---

279. Matt Cane, "What Makes a Power Play Successful?," *Hockey Abstract 2017*, 180.

280. David Staples, "The Eternal Search for the Perfect Hockey Stat, Part 8,348," *Edmonton Journal*, February 13, 2009, http://edmontonjournal.com/sports/hockey/nhl/cult-of-hockey/the-eternal-search-for-the-perfect-hockey-stat-part-8348.

281. Rob Vollman, "Who Is the Luckiest Team?," *Hockey Abstract* (2013), 62–75.

of the coaching staff, particularly when it's broken down by manpower situation, period, or score.

TOP-10 GOALIE MACHINE (T10GM)

In August 2014, Mike Morris introduced the idea of comparing a goalie's save percentage to the average performance of the league's top-10 goalies, rather than compare it to the league average or replacement level.[282]

TOTAL HOCKEY RATING (THOR)

Michael Schuckers and Jim Curro introduced a new catch-all individual player metric in February 2013, the total hockey rating.[283] It assigns values to various events based on their relationship with goals scored and prevented, and then it adjusts for the context in which a player is used, such as manpower situation, zone starts, score effects, and quality of competition and linemates. It was originally measured relative to an average player, but within a couple of months it was changed to a replacement-level player.

TOTAL PERFORMANCE CHARTS

Based on the popularity of player usage charts, I introduced total performance charts in *Hockey Prospectus 2013–14*.[284] They included special teams usage charts, offensive profile charts, and cap efficiency charts.

TOUGH MINUTES

A player usage assignment that usually begins in the defensive zone against top-line competition and/or alongside replacement-level linemates (or some reasonably tough combination thereof) is referred to as playing the tough minutes.

TOUGH START

Being assigned a start behind a tired blue line on the second night of a back-to-back road trip can be a tough game for a goalie, which is why it is usually reserved for the backup, as Tom Awad established in

---

282. Mike Morris, "Jimmy Versus the Top 10 Goalie Machine," *Winging It in Motown* (blog), August 30, 2014, http://www.wingingitinmotown.com/2014/8/30/6086655/jimmy-versus-the-top-10-goalie-machine.

283. Michael Schuckers and James Curro, "Total Hockey Rating (THoR)," Stat Sports Consulting, December 2013, http://statsportsconsulting.com/thor/.

284. Rob Vollman, "Introducing Total Performance Charts," *Hockey Prospectus 2013–14*, x–xv.

Hockey Abstract 2014.[285] (Although a former backup goalie once privately referred to these types of tough starts with a more colourful adjective.)

TRUE PLUS/MINUS

In October 2008, David Staples introduced the concept of a player's true plus/minus as good plays that led directly to goals scored minus bad plays that resulted in goals against. Eventually this idea became the basis of Tarasov numbers.[286]

TRUE SHOOTING PERCENTAGE (TSH%)

While a player's standard shooting percentage is based only on the shots that reached the net or would have reached the net in the opinion of the scorekeeper, true shooting percentage is based on all shot attempts, including those that were blocked or that missed the net. In some circles, it is more commonly known as Corsi shooting percentage (CSh%).

TURNOVERS

To improve the NHL's giveaway statistic, some analysts manually track turnovers, which were defined by Andrew Berkshire in November 2016 as "any play a player makes that ends up with the opposition possessing the puck."[287]

TWO-PERIOD SHOT-FOR PERCENTAGE (2PS%)

Ben Wendorf introduced 2pS% in April 2013 as a historical proxy for modern shot-based metrics, which have only been available since 2007–08. It's the number of shots taken by a team divided by all the shots taken in the game. To account for score effects, it is based only on the first two periods.[288]

ULTIMATE FACEOFF PERCENTAGE (UFO%)

Introduced by Timo Seppa in August 2011, UFO% is a player's regular

285. Tom Awad, "What Makes Good Players Good Part 3, Goaltenders," *Hockey Abstract 2014*, 76–82.

286. David Staples, "Frequently Asked Questions About True Plus/Minus," *Edmonton Journal*, October 14, 2008, http://edmontonjournal.com/sports/hockey/nhl/cult-of-hockey/frequently-asked-questions-about-true-plusminus.

287. Andrew Berkshire, "Giveaways vs Turnovers: How to Find Value in an Unreliable Stat," Sportsnet, November 26, 2016, http://www.sportsnet.ca/hockey/nhl/giveaways-vs-turnovers-find-value-unreliable-stat/.

288. Ben Wendorf, "Possession Corsi Zone Time," *Hockey Visualized* (blog), April 5, 2013, https://hockeyvisualized.com/2013/04/05/possession-corsi-zone-time/.

faceoff winning percentage, but only on the road and at five-on-five. It is adjusted for the average quality of the opposing centres.[289]

USAT

See Fenwick (page 270).

VENUE-ADJUSTED STATISTICS

While the impact of playing at home and away has been studied for decades, venue-adjusted statistics were virtually unheard of until Micah Blake McCurdy included them in his adjusted possession measures in November 2014.[290] That's because, generally, very few players wind up playing significantly more games at home than on the road, except for goalies.

VUKOTA

In 2003, Nate Silver of *Baseball Prospectus* developed the PECOTA system to project a player's performance for the coming season.[291] It inspired similar systems in other sports, including the VUKOTA system developed by Tom Awad in July 2009, which is explained in more detail in *Hockey Prospectus 2010–11*.[292]

WARRIOR CHART

See HERO chart (page 273).

WEIGHTED ENTRIES (WE/60)

In August 2017, Charlie O'Connor first advanced the idea of evaluating a player's ability to gain the zone using his zone entries per 60 minutes weighted by the number of shot attempts generated from each type of entry, such as controlled and uncontrolled.[293]

289. Timo Seppa, "No Advanced Faceoff Metric?," *Hockey Prospectus 2011–12*, 390–394.

290. Micah Blake McCurdy, "Adjusted Possession Measures," *Hockey Graphs* (blog), November 13, 2014, https://hockey-graphs.com/2014/11/13/adjusted-possession-measures/.

291. Nate Silver, "Introducing PECOTA," in *Baseball Prospectus* by Joseph Sheehan, Clay Davenport, and Chris Kahrl (Potomac Books, 2003).

292. Tom Awad, "Introducing VUKOTA," *Hockey Prospectus* (blog), July 20, 2009, https://www.hockeyprospectus.com/puck/unfiltered/?p=68; Tom Awad, "Introducing GVT and VUKOTA," *Hockey Prospectus 2010–11*, vii–x.

293. Charlie O'Connor, "Measuring the Importance of Individual Player Zone Entry Creation," *Hockey Graphs* (blog), August 10, 2017, https://hockey-graphs.com/2017/08/10/measuring-the-importance-of-individual-player-zone-entry-creation.

WEIGHTED POINTS ABOVE REPLACEMENT (WPAR)

A catch-all statistic similar to GVT and WAR, it was introduced by Evolving Wild in August 2017.[294]

WEIGHTED PRIMARY SHOT CONTRIBUTIONS (WPSC)

Just as weighted shots place a value on each shot attempt based on location and other quality factors, weighted primary shot contributions place a weight on all shot attempts and the passes that immediately preceded them. It was introduced by Alan Wells in September 2016.[295]

WEIGHTED SHOTS (WSH)

Weighted shots are quality-adjusted shots and virtually synonymous with expected goals. The term was first used by Graeme Johns in October 2004, as he assigned a weight to each shot equal to the probability that it would result in a goal based on the shot's location, in a fashion similar to Alan Ryder's expected goals.[296] Johns used weighted shots to measure an individual's own shots, but in May 2012, Brian Macdonald explored the other uses of weighted shots, such as for quality-adjusted save percentages like SQNSV or DIGR, catch-all metrics, and quality-adjusted shot-based metric like expected goals. These concepts were re-visited by Tom Tango in November 2014 and followed up by Micah Blake McCurdy and Matt Cane a month later.[297]

WIN THRESHOLD

Win threshold is a team statistic first introduced by Phil Myrland in October 2009 as the minimum save percentage required for an even-goal differential, or win-loss record. The formula is shots against

---

294. EvolvingWild, "Introducing Weighted Points Above Replacement – Part 1," *Hockey Graphs* (blog), August 1, 2017, https://hockey-graphs.com/2017/08/01/introducing-weighted-points-above-replacement-part-1.

295. Alan Wells, "Weighted Shot Rates Based on the Passing Project," *Hockey Graphs* (blog), September 6, 2016, https://hockey-graphs.com/2016/09/06/weighted-shot-rates-based-on-the-passing-project/.

296. Graeme Johns, "Statistical Shot Quality Weighting" (PDF), Hockey Analytics, October 17, 2004, http://hockeyanalytics.com/Research_files/Weighted_Shots.pdf.

297. Brian Macdonald, Craig Lennon, and Rodney Sturdivant, "Evaluating NHL Goalies, Skaters, and Teams Using Weighted Shots" (PDF), Cornell University Library, May 8, 2012, https://arxiv.org/pdf/1205.1746.pdf; Tom Tango, "Introducing Weighted Shots Differential (aka Tango)," *Tango Tiger* (blog), November 30, 2014, http://tangotiger.com/index.php/site/comments/introducing-weighted-shots-differential-aka-tango; Matt Cane, "Score Adjusted Weighted Shots," *Puck Plus Plus* (blog), December 10, 2014, https://puckplusplus.com/2014/12/10/score-adjusted-weighted-shots/.

minus goals for divided by shots against.[298] In April 2016, Nick Mercadante introduced a completely different statistic with the same name to measure how often a goalie plays well enough to steal a game by posting a single-game GSAA of 0.755 or better.[299]

WINS ABOVE REPLACEMENT (WAR)

Although its exact history is hard to unravel, WAR was introduced in baseball based on the work of Bill James in the 1980s, and it is meant to measure all of a player's contributions in terms of wins relative to the contributions of a replacement-level player. While there were versions of that statistic in hockey as early as 2002, the first statistic actually named WAR was developed by Andrew Thomas and Sam Ventura starting in October 2014.[300] Two years later, a revised WAR model was developed by Dawson Sprigings.[301]

WOODMONEY

Introduced by bloggers WoodGuy and GMoney in July 2016, WoodMoney is a quality of competition metric that classifies all opponents into one of three groups based on their scoring, ice time, and shot-based metrics, and it breaks down how players perform against each group.[302]

WOWY (WITH OR WITHOUT YOU)

As best described by David Johnson, who first made WOWY analyses available for hockey in fall 2008, WOWY "looks at pairs

---

298. Phil Myrland, "The Win Threshold," *Brodeur Is a Fraud* (blog), October 10, 2009, http://brodeurisafraud.blogspot.ca/2009/10/win-threshold.html.

299. Nick Mercadante (@NMercad), "Win Threshold% (WT% is % of starts performing well enough in 5v5 Mercad to win avg NHL game) colored by actual win %," Twitter, April 22, 2016, 11:29 p.m., https://twitter.com/nmercad/status/723579523401392128.

300. Andrew Thomas, "The Road to WAR Series (Index)," *War on Ice* (blog), April 8, 2015, http://blog.war-on-ice.com/index.html%3Fp=429.html.

301. Dawson Sprigings, "Testing and Final Remarks," *Hockey Graphs* (blog), October 28, 2016, https://hockey-graphs.com/2016/10/28/testing-and-final-remarks/.

302. Woodguy, "WoodMoney: A New Way to Figure Out Quality of Competition in Order to Analyze NHL Data," *Woodblog*, July 18, 2016, http://becauseoilers.blogspot.ca/2016/07/woodmoney-new-quality-of-competition.html.

of players and how they perform together and how they perform apart."[303]

Z-RATINGS

See KRACH ratings (page 276).

ZONE-ADJUSTED STATISTICS

Since it's obviously easier for a team to generate a shot attempt after an offensive zone faceoff rather than a defensive zone faceoff, it makes sense that teams or players who frequently start in the offensive zone will have better shot-based metrics than those who don't. Based on calculations created by Timothy Barnes in December 2007 (and adjusted by Matt Fenwick in December 2008), Tore Purdy formally defined zone-adjusted statistics in August 2009 by adding or subtracting 0.8 shot attempts for every additional offensive or defensive zone start at even strength.[304] In September 2011, Jared L introduced a version that incorporated neutral zone faceoffs and on-the-fly changes.[305] While keeping the same basic formula as Purdy, several analysts have landed on different estimates for the value of each additional zone start over the years, with the true value believed to fall somewhere between 0.25 and 0.4.[306] In January 2012, David Johnson circumvented this uncertainty by making the adjustment of simply ignoring the first 10 seconds after any faceoff, after which the advantage of starting in the offensive zone disappears.[307] Likewise, Matt Pfeffer introduced his method in March 2014, which involves breaking

---

303. David Johnson, "Hockey Statistics: From Raw Data to WOWY" (presentation at the Ottawa Hockey Analytics Conference, February 7, 2015), http://statsportsconsulting.com/main/wp-content/uploads/JOHNSON_OTTANALYTICS.pdf; David Johnson, "A WOWY Analysis Guide," *Maple Leafs Hot Stove* (blog), September 29, 2014, https://mapleleafshotstove.com/2014/09/29/wowys/.

304. Timothy Barnes, "Fenwick," *Irreverent Oilers Fans* (blog), December 2007, http://vhockey.blogspot.com/2007/12/fenwickd.html (site discontinued); Matt Fenwick, "Flames Game Night," *Battle of Alberta* (blog), December 12, 2008, http://battleofalberta.blogspot.com/2008/12/flames-game-night_12.html; Tore Purdy, "Corsi Corrected for Starting Shift Location," *Objective NHL* (blog), August 9, 2009, http://objectivenhl.blogspot.com/2009/08/corsi-corrected-for-starting-shift.html.

305. Jared L, "On Zone Starts," *Driving the Play* (blog), September 8, 2011, http://drivingplay.blogspot.ca/2011/09/on-zone-starts.html.

306. Eric Tulsky, "Review: Zone Start Adjustments to Shot Differential," *NHL Numbers* (blog), November 5, 2012 http://nhlnumbers.com/2012/11/5/zone-start-adjusted-corsi-corrections-faceoffs.

307. David Johnson, "Adjusting for Zone Starts," *Hockey Analysis* (blog), January 23, 2012, http://hockeyanalysis.com/2012/01/23/adjusting-for-zone-starts/.

down a player's shot-based metrics in shifts started in the defensive, neutral, and offensive zones and then recombining them in a standard league-average ratio.[308]

ZONE ENTRY AND EXIT PROJECTS

Inspired by the work of Lloyd Percival, Anatoli Tarasov tracked his team's success rate in entering the offensive zone as far back as 1969.[309] Geoff Detweiler and Eric Tulsky re-created that work with the Philadelphia Flyers during the 2011–12 season before expanding their efforts league-wide with the Zone Entry Project in June 2012.[310] A year later, Pierce Cunneen expanded this effort to include zone exits.[311] A year later, these efforts would inspire the All Three Zones Project (see page 260).

ZONE ENTRY TO FORMATION RATE (ZEFR)

In April 2016, Arik Parnass introduced ZEFR rate as "a new and better way to evaluate power plays."[312] It uses manually tracked data to calculate the percentage of zone entry attempts on the power play that either result in a scoring chance off the rush or the team getting set up in formation. In August 2017, Matt Cane introduced an estimated version.[313]

ZONE START ADJUSTED SAVE PERCENTAGE

In April 2012, David Johnson observed that the league-average save percentage is 0.979 within 10 seconds of a faceoff and 0.914 otherwise. Since there's a lot of variation in the volume of such shots different goalies

---

308. Matt Pfeffer, "A (Minor) Re-Calculation of Corsi: Adjusting for Zone Starts," *Hockey Prospectus* (blog), March 12, 2014, http://www.hockeyprospectus.com/a-minor-re-calculation-of-corsi-adjusting-for-zone-starts/.

309. David Staples, "The Eternal Search for the Perfect Hockey Stat, Part 8,138," *Edmonton Journal,* February 13, 2009, http://edmontonjournal.com/sports/hockey/nhl/cult-of-hockey/the-eternal-search-for-the-perfect-hockey-stat-part-8348.

310. Eric Tulsky, "Zone Entries: Introduction to a Unique Tracking Project," *NHL Numbers* (blog), June 20, 2012, http://nhlnumbers.com/2012/6/20/zone-entries-introduction-to-a-unique-tracking-project.

311. Pierce Cunneen, "Zone Exit Tracking Project," *Copper & Blue* (blog), June 24, 2013, http://www.coppernblue.com/2013/6/24/4458526/zone-exit-tracking-project.

312. Arik Parnass, "ZEFR Rate: A New and Better Way to Evaluate Power Plays," *Hockey Graphs* (blog), April 18, 2016, https://hockey-graphs.com/2016/04/18/zefr-rate-a-new-and-better-way-to-evaluate-power-plays/.

313. Matt Cane, "What Makes a Power Play Successful?" *Hockey Abstract 2017,* 186.

face, based on how often faceoffs occur in the defensive zone, Johnson developed a zone-adjusted save percentage that ignores such shots.[314]

ZONE START PERCENTAGE (ZS%)

Believed to have been introduced by Timothy Barnes in November 2007, zone start percentage is the number of faceoffs for which a player lined up in the offensive zone divided by the total number of faceoffs in both the offensive and defensive zones. Unless explicitly stated otherwise, this statistic ignores faceoffs in the neutral zone and on-the-fly changes. It is typically calculated at five-on-five only.[315]

ZONE TIME

In the 1997–98 season, the NHL began recording how much of the game was spent in each zone. This was discontinued after the 2001–02 season.

---

314. David Johnson, "Zone Start Effects on Goalie Save Percentage," *Hockey Analysis* (blog), April 26, 2012, http://hockeyanalysis.com/2012/04/26/zone-start-effects-on-goalie-save-percentage/.

315. Timothy Barnes, "Out for the Defensive Zone Draw," *Irreverent Oilers Fans* (blog), November 2007, http://vhockey.blogspot.com/2007/11/out-for-defensive-zone-draw.html.

CONCLUSION

As per tradition, this is the only conclusion you'll find in a *Hockey Abstract* book. Analytics can certainly shed light on an issue, but the great debates never reach a conclusion.

I'd like to thank the many colleagues I consulted and/or who reviewed my work, many of whom can't be named because of their affiliations. Those I can mention include Cole Anderson, Brooke Boyd, Derek Braid, Matt Cane, Petter Carnbro, Matt Coller, Adrian Dater, Drina Drummond, Iain Fyffe, Jim Jamieson, Micah Blake McCurdy, Michael Peterson, Jacquie Pierri, Neil Pierson, Alan Ryder, Timo Seppa, Chris Snow, Tom Tango, Josh Weissbock, and Carolyn Wilke. I'd also like to thank our illustrator, Joshua Smith, for his hilarious cartoons once again.

Remember that all the raw data for the charts and tables can be found on the Hockey Abstract website (hockeyabstract.com), which also includes information about how to contact me. Please don't be shy about reaching out—you are the reason this book exists.

Let the discussions continue.

"We shall not cease from exploration. And the end of all our exploring will be to arrive where we started and know the place for the first time."

— T.S. Eliot.

ABOUT THE AUTHOR

A former member of the Professional Hockey Writers Association, Rob Vollman was first published in the Fall 2001 issue of the *Hockey Research Journal*. He has since co-authored all six *Hockey Prospectus* books and two *McKeen's* magazines. This is the fifth book he has authored as a part of his Bill James–inspired Hockey Abstract series, which includes the highly popular 2016 book, *Stat Shot*. Rob is one of the field's most trusted and entertaining voices, and he has helped bring what was once a niche hobby into the mainstream. He lives in Calgary, Alberta.